Measurement of Mobile Antenna Systems

For a listing of recent titles in the *Artech House Antennas and Propagation Library*,
turn to the back of this book.

Measurement of Mobile Antenna Systems

Hiroyuki Arai

Artech House
Boston • London
www.artechhouse.com

Library of Congress Cataloging-in-Publication Data
Arai, Hiroyuki.
Measurement of mobile antenna systems / Hiroyuki Arai.
p. cm. — (Artech House antennas and propagation library)
Includes bibliographical references and index.
ISBN 1-58053-065-6 (alk. paper)
1. Mobile communication systems. 2. Antennas (Electronics)—Measurement. I. Title. II. Series.

TK6570.M6 A68 2001 00-066373
621.3845—dc21 CIP

British Library Cataloguing in Publication Data
Arai, Hiroyuki
Measurement of mobile antenna systems. — (Artech House antennas and propagation library)
1. Mobile communication systems 2. Antennas (Electronics)—Measurement
I. Title
621.3'845

ISBN 1-58053-065-6

Cover design by Igor Valdman

685 Canton Street
Norwood, MA 02062

International Standard Book Number: 1-58053-065-6
Library of Congress Catalog Card Number: 00-066373

10 9 8 7 6 5 4 3 2 1

To my parents

Contents

Preface

The spread of mobile telephone service has changed the telephone age. In the early days of telephony, in a typical family, a single phone served multiple users. Currently, the movement is toward a personal phone for each individual. In addition, mobile terminals are not used merely as telephones, but also as personal computers (PCs) and personal digital assistants (PDAs), and many other electrical products. The use of electromagnetic waves is mandatory for mobile communication systems because we cannot function efficiently if we are tied to a wire for our communications. An important component of a radio communication system, therefore, is the antenna, which is used for both transmission and reception of the electromagnetic waves. The antenna determines the sensitivity of the receiver and the radiation efficiency of the transmitter. An antenna's exact performance cannot be calculated numerically because the surroundings and free-space radiation conditions affect antenna performance. The recent remarkable progress in the area of computing resources may help at the design stage, but it is still necessary to confirm antenna characteristics by taking actual measurements during development. In the design of a base station antenna and the infrastructure of a mobile communication system, a cut-and-try technique is widely used in addition to computer-aided design. An expert engineer can often complete the antenna development process in the shortest amount of time by using the cut-and-try technique, and this procedure also requires measurements to be made on the antenna.

The purpose of this book is to present measurement methods used on antennas for mobile communication systems. Prior to making antenna measurements, the propagation characteristics of radio waves in the operational area need to be measured as part of the design of the mobile communication system. Although a great deal of data on propagation characteristics has already been obtained by measurements, it is important that further measurements be made to confirm design performance in the real operational environment. Conventional voice communication requires a narrow-frequency bandwidth for its transmission, while high-speed data transmission for the seamless connection with existing cable networks requires a wide-frequency bandwidth. Chapter 1 presents definitions of basic terminology for both narrowband and wideband propagation measurements, and also describes typical measurement parameters and measurement systems.

The most common component of a mobile communication terminal is a portable phone. Chapter 2 deals with the necessary measurement procedures for portable phones and mobile terminal antennas. Following the definitions of measurement parameters, practical measurement procedures that can be carried out in indoor facilities are described. In addition to these measurements, an electromagnetic compatibility (EMC) or electromagnetic emission (EMI) test is a mandatory requirement before releasing the antenna as a commercial product and a typical measurement method used for this purpose is presented. Chapter 2 also describes mobile terminal antenna tests used for product evaluation.

In the most common position in which a handset terminal is used, the antenna characteristics are seriously affected by the proximity of the hand and head of the user. A stable measurement facility that simulates the human body is therefore necessary during the design of the handset antenna so as to take into account the effects of hand and head. For this purpose, a special type of mannequin called a *phantom* is used during measurements. Chapter 3 presents various types of phantoms used in practice and the techniques used to evaluate the influence of the human body on antenna characteristics. Measurement examples are given using a phantom in place of a human operator, and data are presented that compare the results of measurements made using a phantom with those for a human operator. In addition to the measurements of antenna characteristics, Chapter 3 also presents examples of the SAR (specific absorption ratio) regulation standard as a safety standard for electromagnetic wave irradiation of the human body. A detailed definition of SAR and methods for its measurement using a phantom are also presented.

In building a mobile communication network, the base station is the most important part of the infrastructure. A cellular system requires a high-performance base station antenna to increase the capacity of subscribers. Chapter 4 presents a propagation estimation method for base station siting in cellular systems, a method for base station antenna design, and a measurement method for base station antennas. With a high-power input to the base station antenna, the problem of passive intermodulation (PIM) is encountered. As a secondary part of Chapter 4, the theory of PIM and a system used for its measurement are described. Tests that need to be conducted prior to product shipping include electrical, mechanical, and reliability performance tests, which are also described in detail and practical examples of which are given.

In the development stage of mobile terminals, the specifications include a test in the multipath fading environment. This test can easily be confirmed by propagation measurements in a real outdoor environment. However, although this test is required in order to obtain a license, propagation measurements in such an environment are not always stable in time. Two methods can be used to reproduce fading phenomenon for mobile terminal measurements. One is a fading simulator to generate fading waves as an output of RF signals, and the other is the reproduction of artificial fading waves as a summation of spatially distributed standing waves, in a field simulator. Chapter 5 describes the principles and the construction of a fading simulator and a field simulator, which are indispensable for the development of mobile terminals.

This book is devoted to the theory of and techniques used for antenna measurements used in mobile communication systems, with detailed examples given for important topics. The book discusses not only the development stages of antennas, but also the maintenance of base station antennas. The author hopes this book will be of help in the design and development of mobile communication systems.

My sincere thanks are due to the reviewers for their patient and skillful advice. I am deeply grateful to the following persons for their technical information, photos, numerical data, and illustrations: Prof. Koichi Itoh, Chiba University; Prof. Nozomu Ishi, Niigata University; Prof. Hiroaki Takahashi, Musashi Institute of Technology; Prof. Nobuhiro Kuga, Tokyo Polytechnic University; Dr. Kazuhiro Fujimori, Okayama University; Dr. Yoshio Ebine, Ms. Yuki Sugimoto, and Mr. Yasuhiro Oda, NTT DoCoMo Inc.; Mr. Masayuki Nakano, IDO Co.; Mr. Toru Matsuoka, Nihon Dengyo Kosaku, Ltd.; Mr. Hiroshi Sugisawa, NHK Spring Co., Ltd.; and Mr. Kanemi Sasaki, and Mr. Naoki Ohnishi, Kokusai Electric Co., Ltd. The author is also grateful to Dr. Paul Winning and Dr. Koichi Ichige, Yokohama National University, for their revisions to the English text. In particular, I thank Dr. Julie Lancashire, senior commissioning editor for the publisher, for her encouragement and schedule planning throughout this work.

1

Propagation Measurements

This chapter discusses common evaluation procedures for propagation measurements of mobile terminals and base stations. Propagation measurements of mobile terminals involve tests in a real environment, whereas testing of base stations at the observation site generates field profiles. The former are carried out as final tests on newly developed hand phones following checks in an indoor measurement environment. The latter tests the performance of a newly built base station antenna for basic propagation characteristics during the construction of new mobile communication systems. Although the purpose of the two measurements is different, identical procedures are used.

1.1 Overview of Propagation Measurements

Propagation measurements can be divided into narrowband and wideband bandwidths. Narrowband measurements use the envelope of a continuous electromagnetic signal with no modulation. From the recorded level of the electric field strength, the propagation loss curves for macroscopic observations of the site can be evaluated as can the fading characteristics for microscopic observations.

The propagation loss or attenuation between the transmission and reception sites depends on the characteristics of the environment. For the design of a wireless system, the transmission power and antenna gain must take this attenuation into account. If the system has a transmission power of P_t, a transmission antenna gain G_t, and a reception antenna gain G_r, then the received power P_r can be obtained by including the propagation loss L as:

$$P_r = LG_rG_tP_t \qquad (1.1)$$

When the signal travels only by way of the direct path, of length d, between the transmitter and the receiver, L is given by:

$$L = \left(\frac{\lambda}{4\pi d}\right)^2 \tag{1.2}$$

and is called the free-space propagation loss factor. In real environments, very often several signal propagation paths exist other than the direct one. Interference between the signals along the various paths causes severe field deviation and multipath fading. The measured data include this fading and often prove to be significantly larger than the free-space propagation loss. A propagation loss measurement of the site includes both of these factors.

Propagation loss is measured within a range from a few hundred meters to tens of thousands of kilometers. If, however, the fading path is less than a half wavelength in free space, the received signal level will frequently drop. If a wireless communication system is to overcome this fading phenomenon, higher power margins must be used to guarantee system quality. Commonly, in terrestrial mobile communication systems, a fading margin of more than 20 dB is used. A knowledge of the fading characteristics of a site allows a suitable fading margin to be selected for the system design. The first part of this chapter describes narrowband propagation measurements.

Another important class of propagation measurements is the so-called "wideband measurement." Recent mobile communication systems use ever increasing, higher speed data transmission rates. The delay profile of the site limits the maximum data transmission rate of the system and requires the measurement of the delay profile of the transmitted signal. When the delay profile is known in advance, a circuit called an equalizer can be used to cancel the unwanted signal delay. In addition, a receiver called a rake receiver can be used to add several delayed signals to the direct propagation path signal and, hence, increase the effective received signal level. Thus, this factor requires a delay profile measurement to be performed for wideband communication systems.

The first part of the chapter also includes definitions of basic terminology for narrowband propagation measurements, such as propagation loss, cross-polarization, and diversity. The second part describes wideband measurements. The final section presents typical measurement systems for both narrowband and wideband propagation measurements, and also describes the parameters measured.

1.2 Field Profile Measurements

Measurement parameters of interest in a narrowband system are the propagation loss and the fading characteristics. To find the electric field strength over an area of more than a few kilometers from the base station, a propagation loss measurement is made. In contrast, if the area of measurement has a radius of only a few meters to a few tens of meters centered on the base station, then only the fading characteristics are determined. For the latter measurement, the average received electric field strength can be regarded as almost constant.

Outdoor propagation environments are roughly classified into suburban areas and city areas from the viewpoint of the fading characteristics. Normally, there are few tall buildings in a suburban environment and the propagation path consists of a strong direct path signal from the base station plus an array of scattered waves at the observation point. In this case, the fading characteristics are categorized as Rician fading. Inside a city, however, the scattering and reflections of the signal from the base station from multiple locations eliminate the direct signal. For this situation almost the same level of multipath signals from many directions is received at the receiving point. The fading characteristics in this situation are classified as Rayleigh fading, which is discussed in more detail in Section 1.2.3. To test mobile communications equipment, a typical propagation environment should be chosen for the measurement site. To find a typical measurement environment, it is necessary to select the propagation site before field testing.

1.2.1 Site Selection

The environment characteristics close to the observation point can seriously affect the measurements. Sites discussed here are classified in terms of fading characteristics, field strength, and cross-polarization ratio. The simplest way to classify a measurement site is to determine whether or not the transmitting station is directly observable from the observation point. If it can be directly observed, it is called a line-of-sight (LOS) measurement site; if not, is it called a non-line-of-sight (NLOS) site. This method of categorization is based on the geometrical theory of optics. The dominant fading characteristics for LOS and NLOS sites are called the Rician and Rayleigh distributions, respectively. Fading characteristics are discussed in detail later, however, we can state in general that Rayleigh fading can be observed inside a city, whereas Rician fading tends to be found in the suburbs [1–3].

When the measurement site is categorized in terms of the field strength, there are three types of site classifications. The average field strength level defined later is roughly divided into strong, medium, and weak regimes. The electric field strength discussed here is at the reception antenna output. The electric field strength level classification greatly depends on the operating frequency and system design, and the strong field region can be classified as a mean field strength level of more than 40 dBμV. The medium field range is then 20 to 40 dBμV and the weak field range is less than 20 dBμV for current mobile communication systems operating in the UHF band (300 MHz to 3 GHz). The base noise level is classified as being around 0 dBμV.

1.2.2 Attenuation Coefficient Measurements

When the electric field strength leaving a transmitting station is measured against the range, r, from the transmitting antenna, it is found that the amplitude decreases as $1/r$ for the condition where there are no reflected or diffracted waves present. For practical measurement of the electric field, where the field is measured in units of decibels, the measured data decrease as $1/r^2$. Figure 1.1 shows an example of the measured electric field in the UHF band within the range of a few meters to a few kilometers. The measured electric field does not follow a $1/r^2$ decay law; instead, in general, it follows a $1/r^n$ ($n > 2$) decay law. This approximation is quite rough and the measured data can be quite scattered. This macroscopic data treatment is commonly used for the design of a base station coverage area. The attenuation coefficient is a basic evaluation factor for propagation characteristic measurement. The received signal levels expressed in dBμV are plotted as a function of distance from the transmitting station. The measured data show significant scattering, so the data decay profile is found using a statistical procedure.

Another propagation measurement can be carried out near the transmitting base station for the purpose of indoor propagation measurements for micro- or picocellular system design. Figure 1.2 plots the results of an example of a short-range measurement of the received electric field in the vicinity of the transmitting station. As shown in the figure, severe fading can be observed. To find the average signal level at the measurement site, a short-term median is used. The definition of "short term" is not clear. When the total measurement length is a few hundred meters, the median value of the collected data in a few meters is used as the short-term median. However, for measurements of more than tens of kilometers, the short term is defined as a few kilometers. The short term is classed as being roughly 1/100 of the total measurement length.

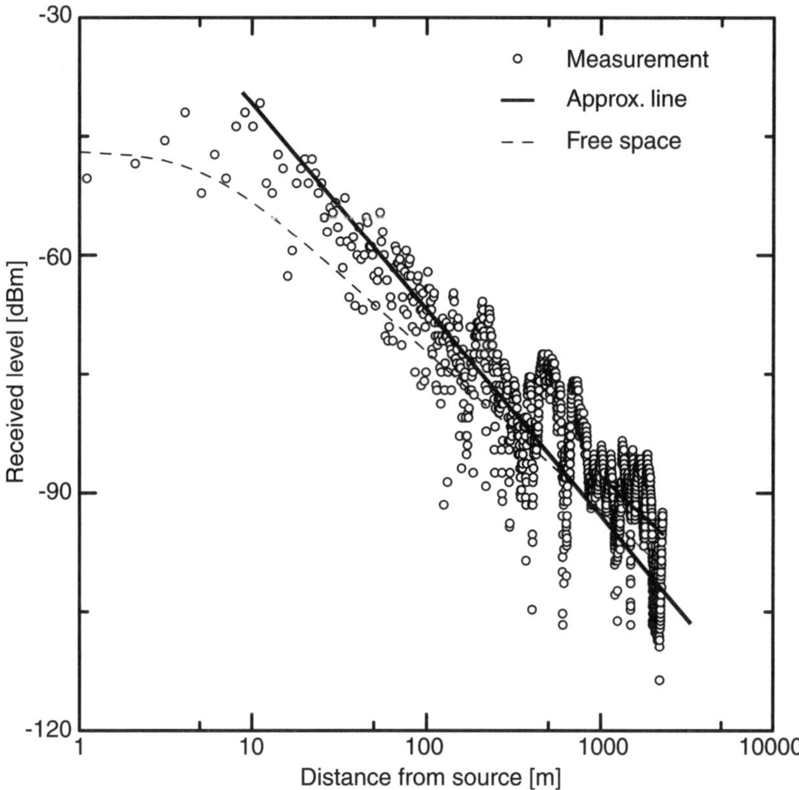

Figure 1.1 Long-range propagation measurement.

A communications receiver detects the free-space radio frequency (RF) signal at intermediate frequencies (IF) using an envelope detector. When seeking to find the average received signal level from the measured data (normally recorded in units of decibels), it is important to take into consideration the method of calculation used. The measured data must first be transformed into linear units before the mean value of the data is sought. A square law detector is used to pick up the electric field but has input–output characteristics that do not always exactly follow the square curve law. If calibration of the detector is carried out for every measurement, the measurement process can prove time-consuming and cumbersome. To exclude such a procedure, a median is used as an average value. The median for received data consisting of a total number of N data points can be estimated using the following procedure. The data are sorted by amplitude with the $(N/2)$th amplitude chosen as the median. The median value eliminates the deep fading in the

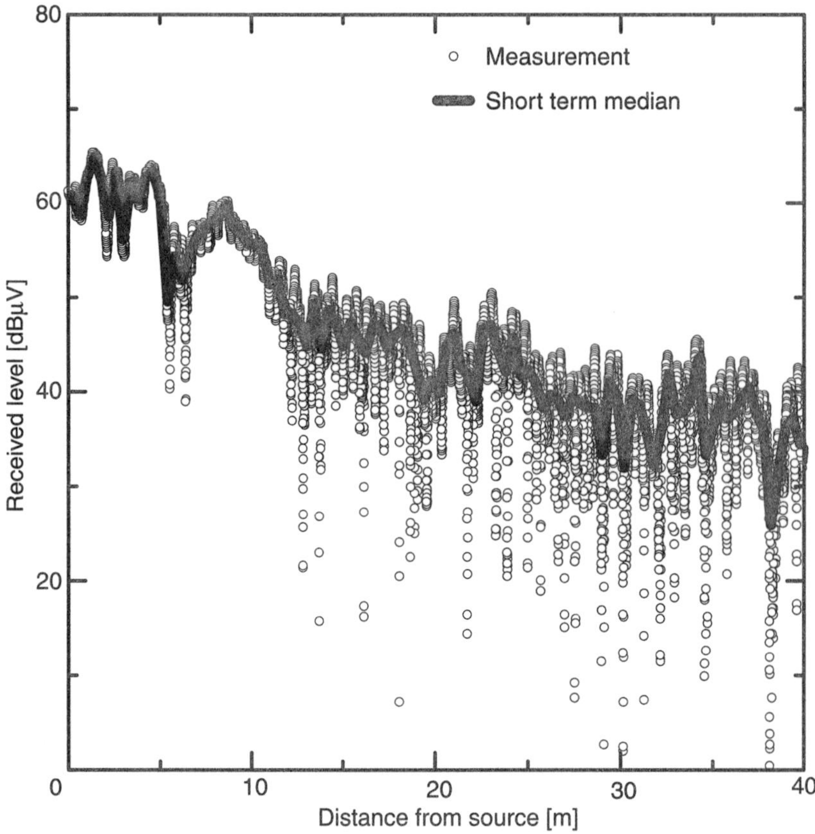

Figure 1.2 Short-range propagation measurement.

measurement and gives an average value for the measurement site. In addition, the cell edge is defined by the level of the median of the downlink (transmitter to receiver) at the mobile terminal in practical cellular systems. The critical value is set at between 10 and 20 dBμV, depending on the system design.

1.2.3 Fading Structure

Electromagnetic waves radiated from a transmitting station reach the observation point by multiple reflections and diffractions of the original wave and these cause multipath fading. The fading structure is modeled here from a statistical point of view. The signals arriving at the observation point are concentrated in a horizontal plane when the transmitting station is located far from the point of observation. Figure 1.3 shows the case for which the ob-

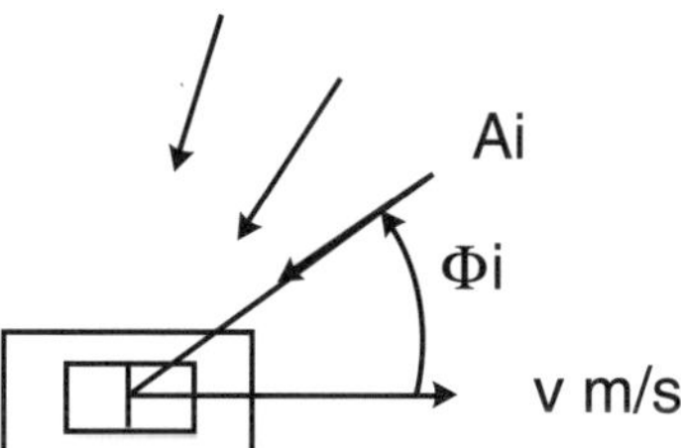

Figure 1.3 Mobile terminal signal reception parameters.

servation point is moving along the x axis with a velocity of v meters per second (m/s). For this situation N waves arrive from the ϕ_i direction, each with an amplitude A_i and phase ζ_i. The time-varying electric field $E_r(t)$ at the observation point is expressed in terms of the angular frequency, ω_c, as:

$$E_r(t) = \sum_{i=1}^{N} A_i \cos\left\{\omega_c t + \zeta_i + 2\pi\frac{v}{\lambda}\, t \cos\phi_i\right\} \tag{1.3}$$

The last term in (1.3) is the frequency shift due to the Doppler effect caused by the movement of the observation point. Using standard trigonometric expressions to expand the above equation, we obtain:

$$\begin{aligned} E_r(t) &= \sum_{i=1}^{N} A_i \cos\psi_i \cos\omega_c t - \sum_{i=1}^{N} A_i \sin\psi_i \sin\omega_c t, \\ \psi_i &= \zeta_i + 2\pi\frac{v}{\lambda}\, t \cos\phi_i \end{aligned} \tag{1.4}$$

When the amplitudes of the individual waves, A_i, are almost the same and the phase ζ_i changes randomly, the coefficients of $\cos\omega_c t$ and $\sin\omega_c t$ are given by the Gaussian distributions $x(t)$ and $y(t)$. The mean values of both $x(t)$ and $y(t)$ are 0, with the same variance, σ, and E_r thus becomes:

$$E_r(t) = x(t)\cos\omega_c t - y(t)\sin\omega_c t \tag{1.5}$$

where $x(t)$ and $y(t)$ are expressed by the following distribution functions:

$$x(t) = \frac{1}{\sqrt{2\pi\sigma}}\exp\left(-\frac{x^2}{2\sigma^2}\right), \quad y(t) = \frac{1}{\sqrt{2\pi\sigma}}\exp\left(-\frac{y^2}{2\sigma^2}\right) \tag{1.6}$$

Then, the Gaussian distributions $x(t)$ and $y(t)$ are replaced by the following probability density function:

$$p(x,y) = \frac{1}{2\pi\sigma^2} \exp\left(-\frac{x^2+y^2}{2\sigma^2}\right) \tag{1.7}$$

where the amplitude of the total electric field strength is denoted by $R = \sqrt{x^2+y^2}$, and the amplitude probability in the range R to $R+dR$ is obtained by integrating (1.7) over the region $0 \le \theta \le 2\pi$ after employing the transformations $x = R\cos\theta$ and $y = R\sin\theta$. Then,

$$p(R) = \frac{R}{\sigma^2} \exp\left(-\frac{R^2}{2\sigma^2}\right) \tag{1.8}$$

This function is known as the Rayleigh distribution and is shown in Figure 1.4.

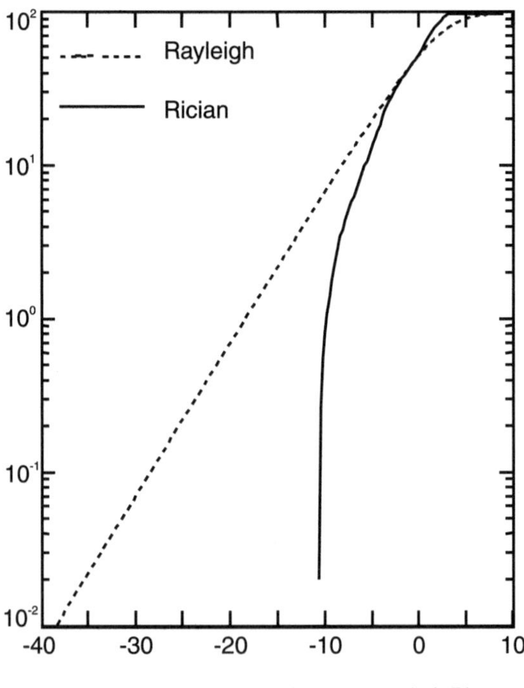

Figure 1.4 Cumulative probability distribution.

The Rayleigh distribution results from the summation of independent waves arriving with almost the same amplitude. On the other hand, if the signal is composed of one strong wave R_0 and several weak diffracted waves, then the received signal is given by a Gaussian distribution with independent variables $X = x + R_0$ and $Y = y$ and the following probability density function:

$$p(X,Y) = \frac{1}{2\pi\sigma^2} \exp\left(-\frac{(X - R_0)^2 + y^2}{2\sigma^2} \right) \tag{1.9}$$

After transformation, $p(R)$ is given by an integral in terms of θ as:

$$p(R) = \frac{1}{\sigma^2} \exp\left(-\frac{(X - R_0)^2 + y^2}{2\sigma^2} \right) I_0\left(\frac{R_0 R}{\sigma^2} \right) \tag{1.10}$$

where I_0 is a zero-order modified Bessel function. This function is known as the Rician distribution.

1.2.4 Cumulative Probability

The preceding section defined the fading structure of a mobile propagation environment using the probability density function. From a practical point of view, it is useful to express the received field strength as a probability function. This is because some amplitude level of R often appears throughout the measured region, and the probability is useful to judge whether the received signal strength is less than a critical level or not. The probability density function, $p(R)$, of the Rayleigh distribution can be rewritten as follows by using the mean value of the square amplitude $\nu = 2\sigma^2$:

$$p(R) = \frac{2R}{\nu} \exp\left(-\frac{R^2}{\nu} \right) \tag{1.11}$$

Then, the probability of a received signal level less than R is given as:

$$P(R) = \int_0^R p(x)\, dx = 1 - \exp\left(-\frac{R^2}{\nu} \right) \tag{1.12}$$

In mobile propagation measurements, the median of the square of the amplitude of field strength is used as the average level for short-range measurements. The above probability can then be rewritten in terms of the amplitude level R_m for $p(R_m) = 50\%$ as:

$$P(R) = 1 - \exp\left\{ -\ln 2 \left(\frac{R}{R_m} \right)^2 \right\}, \quad R_m = \sqrt{\nu \ln 2} \tag{1.13}$$

This probability, normalized by R_m, is shown in Figure 1.4, and is called the cumulative probability distribution. For Rician fading, the cumulative probability function is given as:

$$p(R) = \frac{2R}{\nu} \exp\left(-\frac{R^2}{\nu} - \gamma \right) I_0 \left(\frac{2R\sqrt{\gamma}}{\nu} \right) \tag{1.14}$$

where $\nu/2$ is the total received power of the weak diffracted waves, and γ is the ratio between the power level of the strong direct wave and $\nu/2$. The Rician distribution is also shown in Figure 1.4, where its fading depth is decreased by a factor γ.

1.2.5 Cross-Polarization

The electromagnetic wave radiated from the base station propagates in free space and consists of electric and magnetic field components that are orthogonal to each other. The propagation characteristics are mainly described by the electric field component. The polarization describes the direction of the wave electric field component.

In the early days, mobile communication systems used vertical polarization in the VHF (30- to 300-MHz) band because the horizontal polarization component becomes very weak at a height of 1 to 2m above the mobile station ground level. Although recent terrestrial mobile communication systems in the UHF band (300 MHz to 3 GHz) transmit vertically polarized waves from the base station, many obstacles in the propagation path reflect or diffract the transmitted waves and excite a proportion of the horizontally polarized component. It is therefore important to measure the cross-polarization ratio (XPR) at a mobile terminal in advance of the propagation measurement. The XPR is defined as the ratio between the vertically and the horizontally polarized components:

$$\text{XPR} = \frac{P_v}{P_h} \tag{1.15}$$

where P_v and P_h are the short-range medians for the vertically and horizontally polarized components, respectively, at the measured site. In the ideal case, the radiation pattern of the measuring antenna should be identical in both planes of measurement (vertical and horizontal). However, in real XPR measurements it is often convenient to use a standard half-wave dipole antenna as the measuring antenna. Figure 1.5 shows typical XPR measurement results around a tall building in the city and in the corridor between offices on the seventh floor of a building. The XPR is high in the LOS region and is degraded in the shadow of the building. The XPR was found to be around 9 dB in a suburban area and from 3 to 6 dB in the city. The XPR decreased to 0 dB at sites surrounded by tall buildings [4].

1.3 Diversity Measurements

Multipath fading at the mobile receiver varies with the sum of the electric field components incident on the receiving antenna from various directions

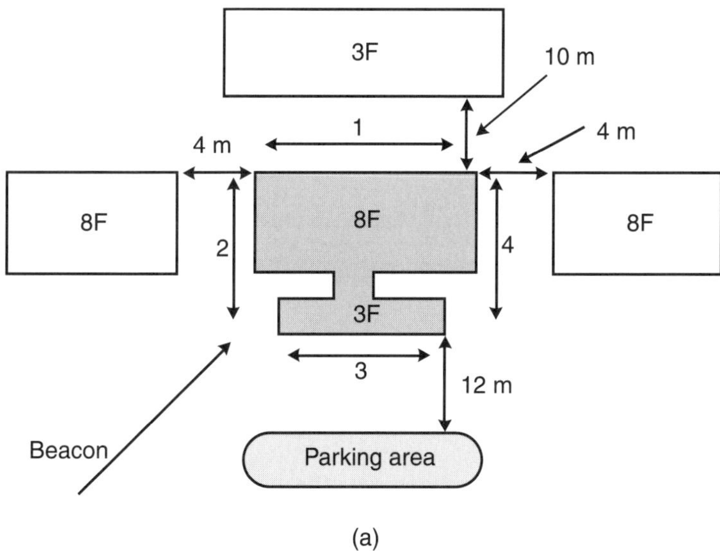

Figure 1.5 (a) Outdoor environment, (b) indoor environment, and (c) measured vertical (V-pol) and horizontal (H-pol) component (dBμV) and cross-polarization ratio (dB).

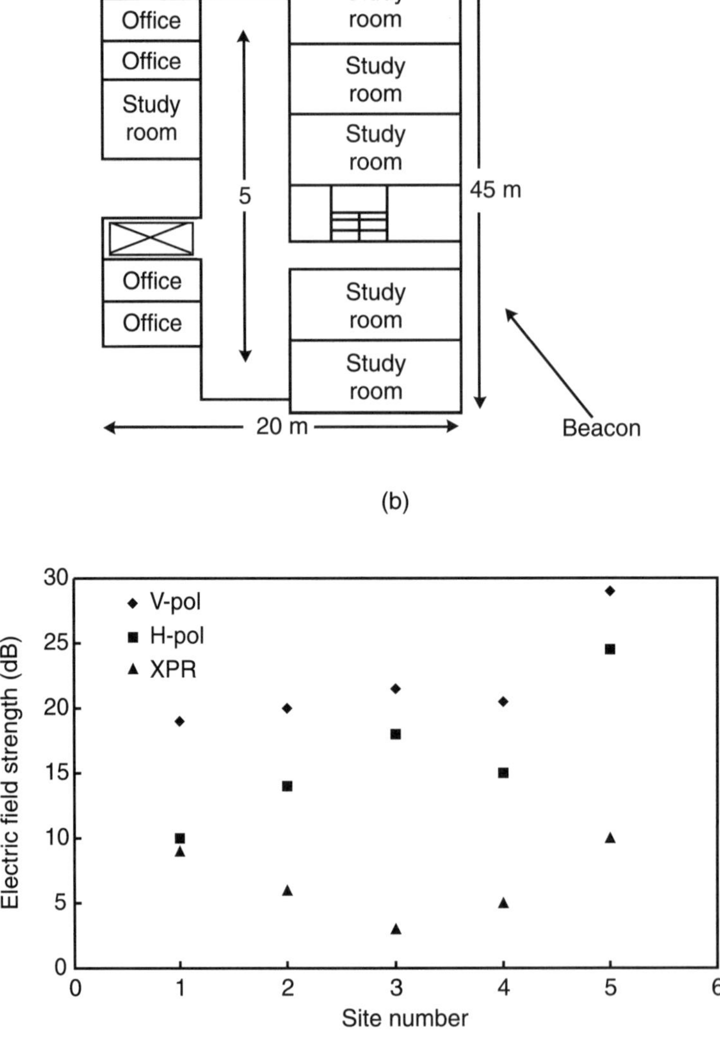

Figure 1.5 Continued.

and can range from about 20 to 40 dB. One technique to help reduce the influence of deep fading is diversity reception. This technique uses several simultaneously received independent signals and synthesizes these signals to compensate for the fading. In this section, diversity schemes employing several input antennas are discussed as well as suitable measurement techniques.

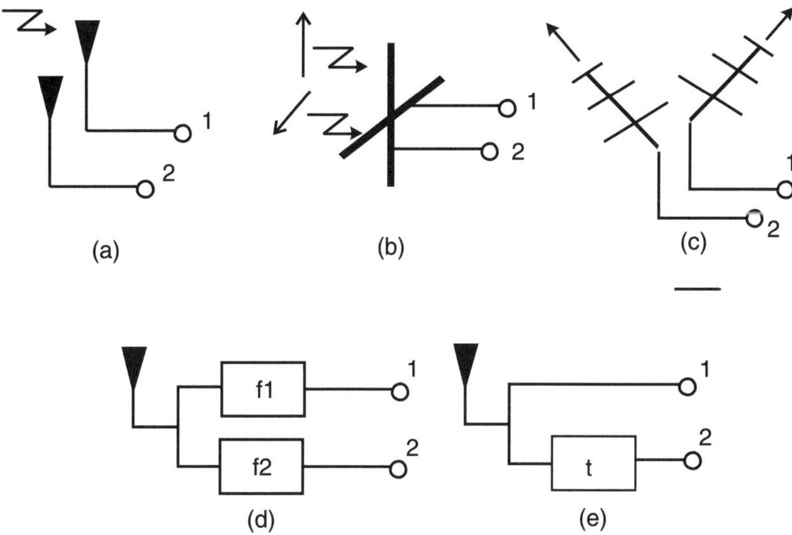

Figure 1.6 Various diversity reception schemes: (a) space diversity, (b) polarization diversity, (c) antenna pattern diversity, (d) frequency diversity, and (e) time diversity.

1.3.1 Diversity Schemes

Diversity reception is used for mobile terminals and base stations to help compensate for very weak signal reception caused by multipath fading. The same increase in communication performance is also expected for the base station. However, it serves an additional purpose. The downlink signal from the base station to the mobile terminal is stronger than the uplink from the mobile terminal to the base station. The base station transmits tens of watts, whereas the mobile terminal radiated power can be less than 1W in current cellular systems. The main purpose of using diversity reception at the base station, therefore, is to compensate for the difference in signal levels between the uplink and the downlink.

Diversity schemes may be categorized into three major types: antenna (or space), frequency, and time, as shown in Figure 1.6 [5]. The latter two methods depend on the type of communication system, whereas the antenna diversity method is independent of the system. This section describes antenna diversity in detail. Antenna diversity techniques use several methods to help reduce fading. One method uses several independent signals from different antennas placed some distance apart. Another method is termed polarization diversity and uses the normal and cross-polarization components of the received signal. Yet another technique, called antenna pattern diversity, utilizes the signals from antennas with nonoverlapping antenna radiation patterns. Antenna pattern diversity is also called angular diversity and has been used for overseas wireless communications in the HF band (3–30 MHz) to eliminate the fading caused by time variations from signals reflected from the ionosphere.

For the base station antenna of a cellular system, space diversity systems are widely used. However, the antenna gain of the base station is approximately 10 dB or more and this requires antennas to be spaced more than 10 to 20 wavelengths apart in order to decrease the mutual interference between the two antennas. Recently, the dramatic increase in the number of subscribers has forced communications companies to install a large number of base station antennas. Polarization diversity schemes using an inclined polarization scheme of ±45 degrees have been adopted in Europe and America to obtain a stable mobile communications uplink. In Japan, a polarization scheme utilizing the horizontal and vertical polarization components has been used because the handheld mobile phone terminal is often used at a tilted angle [6].

One type of diversity antenna that helps to reduce the number of incoming waves required to be received by the base station is the so-called "sector antenna." This antenna divides the coverage area into three or more sectors. In high-speed digital data transmission the sector antenna helps to eliminate the delayed signal due to multiple reflections and may be used for applications such as wireless local-area networks (LANs) or mobile communication systems.

1.3.2 Correlation Histogram and Coefficient

The performance of diversity antennas is evaluated by measuring the amplitude and phase components of the radiation pattern, or by testing the antenna in a practical environment. These radiation patterns are hereafter referred to as complex radiation patterns. In both cases, the evaluation factor is the correlation coefficient between different diversity branches of the antenna. Generally, the term *diversity branch* denotes the multiple output from the antenna after the modulated signal is detected, but in this section it refers to the different antenna output ports. When there are two output branches for a particular diversity antenna, the correlation coefficient of the detected signal envelope of e_1 and e_2 is defined as:

$$\rho = \frac{\frac{1}{2}\left\langle \left(e_1 - \langle e_1 \rangle\right)^* \left(e_2 - \langle e_2 \rangle\right) \right\rangle}{\frac{1}{2}\sqrt{\left\langle \left(e_1 - \langle e_1 \rangle\right)^2 \left(e_2 - \langle e_2 \rangle\right)^2 \right\rangle}} \tag{1.16}$$

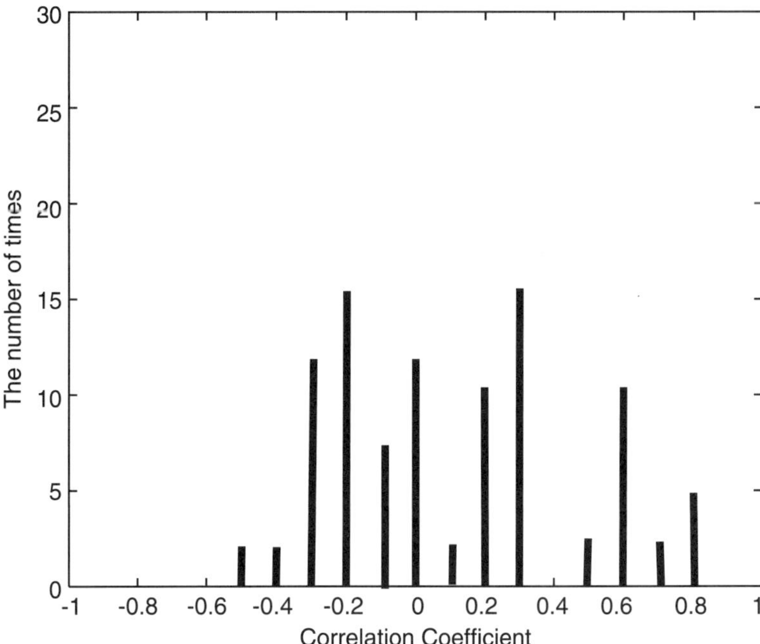

Figure 1.7 A histogram of the correlation coefficient.

where <*x*> denotes the mean value of the measured parameter *x*, and $(y)^*$ denotes the complex conjugate of the difference of the measured parameter (y). In propagation measurements the correlation coefficient is calculated every 100 or 200 samples, and a correlation histogram of the form shown in Figure 1.7 is built up.

In practical diversity measurements, more than tens of thousands of data samples are collected in one measurement. Thus, for a single measurement, hundreds of correlation coefficients must be calculated. Due to the large number of data samples that must be collected, it is best if the correlation coefficient is calculated in the region where the average received electric field level is almost constant, thus reducing the total number of calculations. The correlation coefficient is used for the macroscopic evaluation of diversity antennas.

1.3.3 Diversity Gain

The effectiveness of diversity reception is evaluated by the correlation coefficient presented in the previous section. Another evaluation factor is the diversity gain. Using diversity reception in a multipath propagation environment reduces the fading depth of the received signal. In a propagation environment exhibiting Rayleigh distribution fading, the cumulative probability distribution of the sum of the received signals from N branches of the diversity antenna is given by the average carrier-to-noise ratio (CNR), γ/Γ. Then:

$$P(\gamma) = \left\{1 - \exp\left(\frac{\gamma}{\Gamma}\right)^{n}\right\} \tag{1.17}$$

where selection diversity is assumed after detection of the received signals and the correlation coefficient between branches is zero. In selection diversity, selecting the branch with highest power combines the data sequences representing the envelope power received by each branch. Figure 1.8 shows an

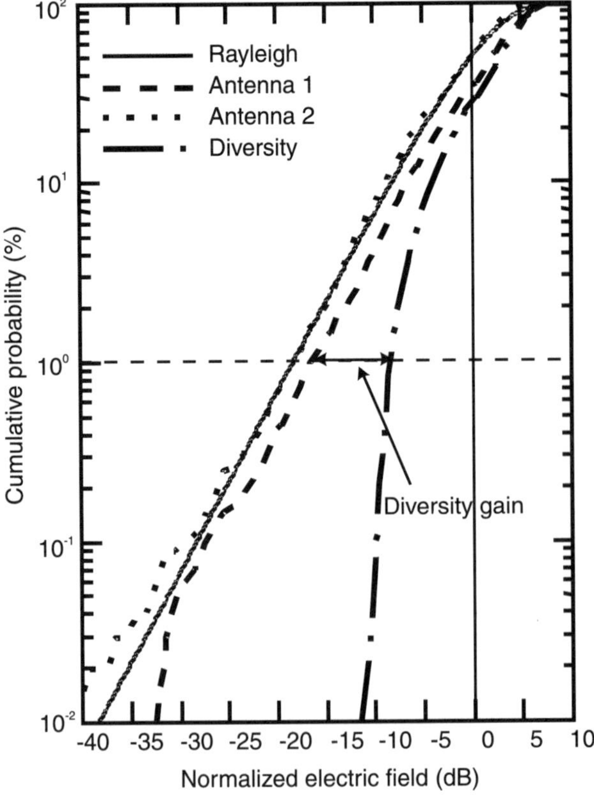

Figure 1.8 Definition of diversity gain.

example of the cumulative probability distribution. In this figure the diversity gain is defined by the difference of the reception level at $P(\gamma) = 1\%$ between curves $n = 1$ and n.

Diversity gain indicates the degree by which diversity reception reduces the fading depth. To evaluate the selection process statistically from the measured propagation data, the envelope of the received signal for each diversity branch is combined by selecting the branch with the highest level of $\max(E_i^1, E_i^2)$. In this case, E_i^1 and E_i^2 represent the received signal levels after the process of envelope detection by branches 1 and 2 of the diversity antenna and i represents the discrete time when the sample was taken.

1.4 Delay Profile Measurements

In wireless high-speed digital transmissions, the desired symbol data can be corrupted by interference effects caused by the interaction of the directly transmitted signal with a reflected path signal. When the data speed is increased, the previously transmitted symbol data can arrive at the receiving point simultaneously with currently transmitted symbol data. The use of diversity reception, adaptive antenna pattern shaping, or implementation of a signal processing system may eliminate these undesired signals. To select an appropriate method for designing a communication system, the characteristics of the delayed signal must be measured in advance at several sites. The evaluation factors needed to measure delayed signals are defined in the following section.

1.4.1 Delay Profile and Delay Spread

For the delayed signal measurement, the base station transmits a pulse signal and the mobile terminal receives the echo. Figure 1.9(a) shows an example of an actual delayed signal profile received at a fixed observation point. To make it easier to understand the definition of the delay profile, the actual received signal shown in Figure 1.9(a) is modeled by an ideal received signal shown in Figure 1.9(b). The distribution function of the received signal strength, as a function of time t, is denoted by the function $f(t)$. Thus, the received power P_m, which includes the delayed pulse, is given by:

$$P_m = \int_{t_0}^{t_3} f(t)\, dt \qquad (1.18)$$

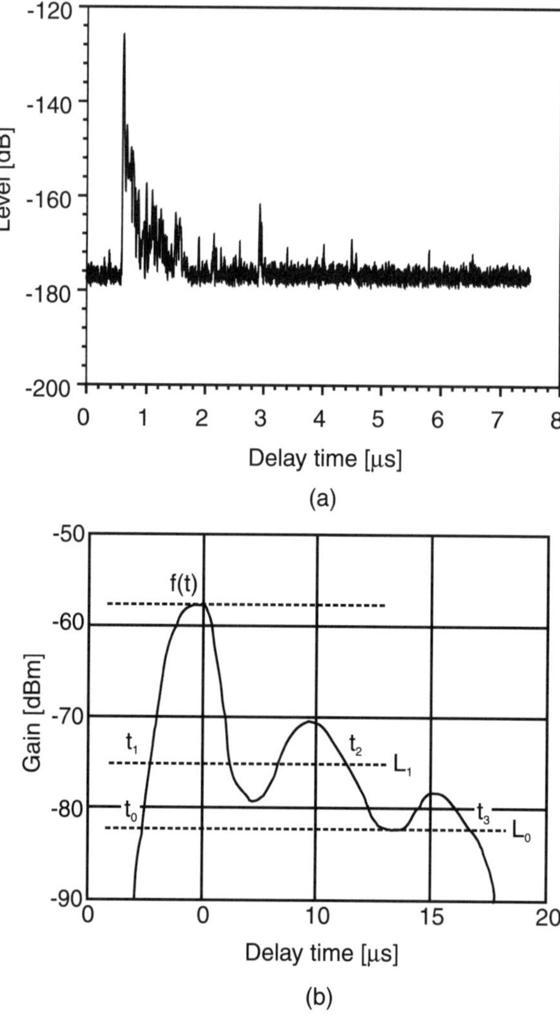

Figure 1.9 Received signal echo profile: (a) measured delay profile and (b) parameters for delay profile.

where t_0 is the time at which $f(t)$ first crosses the noise level L_0, and t_3 is the time at which $f(t)$ next crosses the noise level L_0. A theoretical definition of t_0 can be calculated using the direct LOS distance from the transmitter to the receiver. However, a definition also including the noise level L_0 is useful for practical evaluation purposes. A second evaluation factor T_D is defined as the mean of the delay time:

$$T_D = \frac{1}{P_m}\int_{t_0}^{t_3} (t - t_0) f(t)\, dt \tag{1.19}$$

from which the delay spread or standard deviation of $f(t)$ can be defined as follows:

$$P_m = \sqrt{\frac{1}{P_m}\int_{t_0}^{t_3} t^2 f(t)\, dt - T_D^2} \tag{1.20}$$

1.5 Propagation Measurement System

This section describes the propagation measurement system used to evaluate the propagation loss factor at a site and also the diversity performance of the system. For the evaluation of both factors, the measurement system is essentially the same except for the multiple receiving antennas required for the diversity measurement. The system performance is, however, heavily dependent on the linearity and response time of the detector and these topics are dealt with later in this section. In the final parts of this section, a low-cost multichannel measuring technique that utilizes antenna switching is presented.

1.5.1 Measurement Systems

Measurement systems can be classified into four types. The most basic system measures the propagation loss factor for both short and long ranges to find the site attenuation coefficient. A second type of measurement system is used to investigate the fading structure and diversity performance of the antenna used in narrowband mobile communication systems. A third type of measurement system finds the delay spread for evaluation of the wideband characteristics of the system. Last, for maintenance and field checking of cellular systems currently in service, fully automated field measurement systems are required. This section describes appropriate measurement systems for each purpose.

1.5.1.1 Long-Range and Short-Range Propagation Loss Measurements

Figure 1.10 illustrates a standard measurement system. The system consists of a signal generator connected to a low-gain antenna and a receiver with a

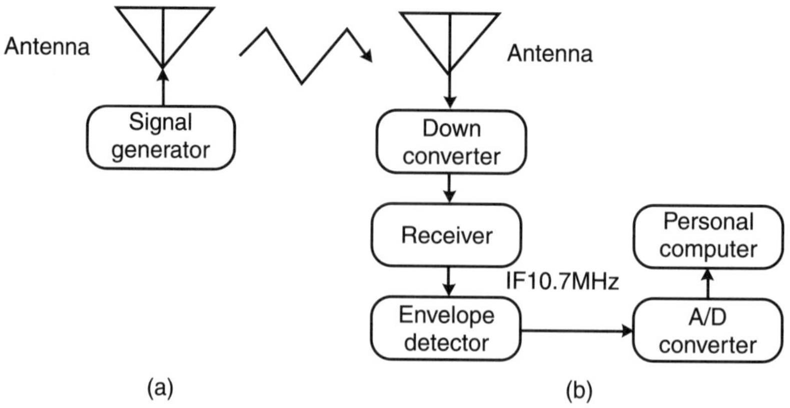

Figure 1.10 A standard measurement system: (a) transmitting system and (b) receiving system.

standard antenna. In this case, the standard antenna is a half-wavelength dipole antenna with a known radiation pattern and antenna gain. The level of the envelope signal at the receiver output after detection is sampled using an analog-to-digital converter (ADC). For long-range measurements, the sampling rate of the ADC is chosen to sample the data at less than a quarter-wavelength spacing. This is because the period of the standing wave created between the transmitting and receiving antenna has a period of one-half wavelength, as shown in Figure 1.11.

Thus, the sampling frequency f_s is determined by the relation $f_s = 4v/\lambda$ Hz, where the receiving system is moving at a rate of v m/s, and λ is the operating wavelength of the signal generator. To record the fading structure in detail without using any interpolation techniques, the sampling spacing should be less than $\lambda/20$ for short-range measurements [7]. Finely spaced

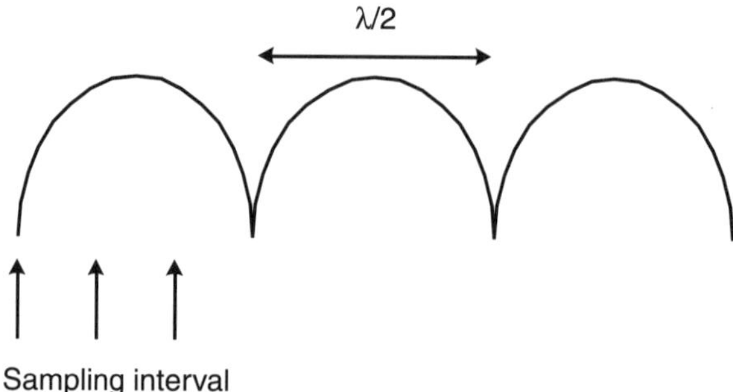

Figure 1.11 The receiving signal standing wave and the sampling interval.

sampling is required to obtain more than a –30-dB fading depth accuracy in the measured Rayleigh distribution data.

1.5.1.2 Diversity Performance Measurements

The only difference between the measurement system used in Section 1.5.1.1 and that used for diversity performance evaluation is the multichannel receiver and multichannel ADC shown in Figure 1.12. For a two-branch diversity system, two identical receiving subsystems are required for simultaneous measurement of the received signals from the two antennas.

A spectrum analyzer can often be used as a receiver for the measurement system, but it does experience large electric power consumption, making it rather inconvenient for portable measurement systems. A commercially available wideband receiver can also be used for the same purpose. The two independent receiver subsystems should also be calibrated using a known power level standard as an input to each channel. A discrepancy of 1 to 2 dB often occurs between the two receivers, which has to be calibrated. When the mobile terminal moves at a speed of 80 km/h, the required sampling rate is 0.68 ms in order to satisfy the $\lambda/20$ sampling interval at 1 GHz. A multichannel ADC usually has a sampling rate of less than a few microseconds and therefore does not affect the overall sampling rate of the system.

To simplify the measurement system, one receiver can be used with multiple antennas. To perform the measurement, the technique of switching between different antennas at a specified switching rate is used. The specific switching rate requirements for such a system are discussed later.

1.5.1.3 Delay Profile Measurements

The delay profile measurement system is shown in Figure 1.13. To perform this type of measurement, a precise timing signal is essential for both the

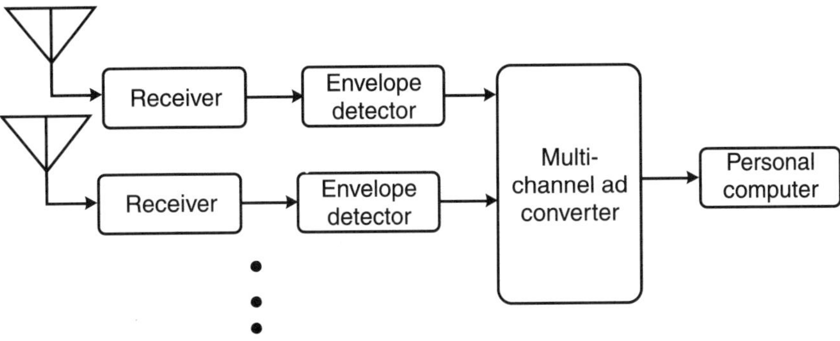

Figure 1.12 Multichannel receiver system.

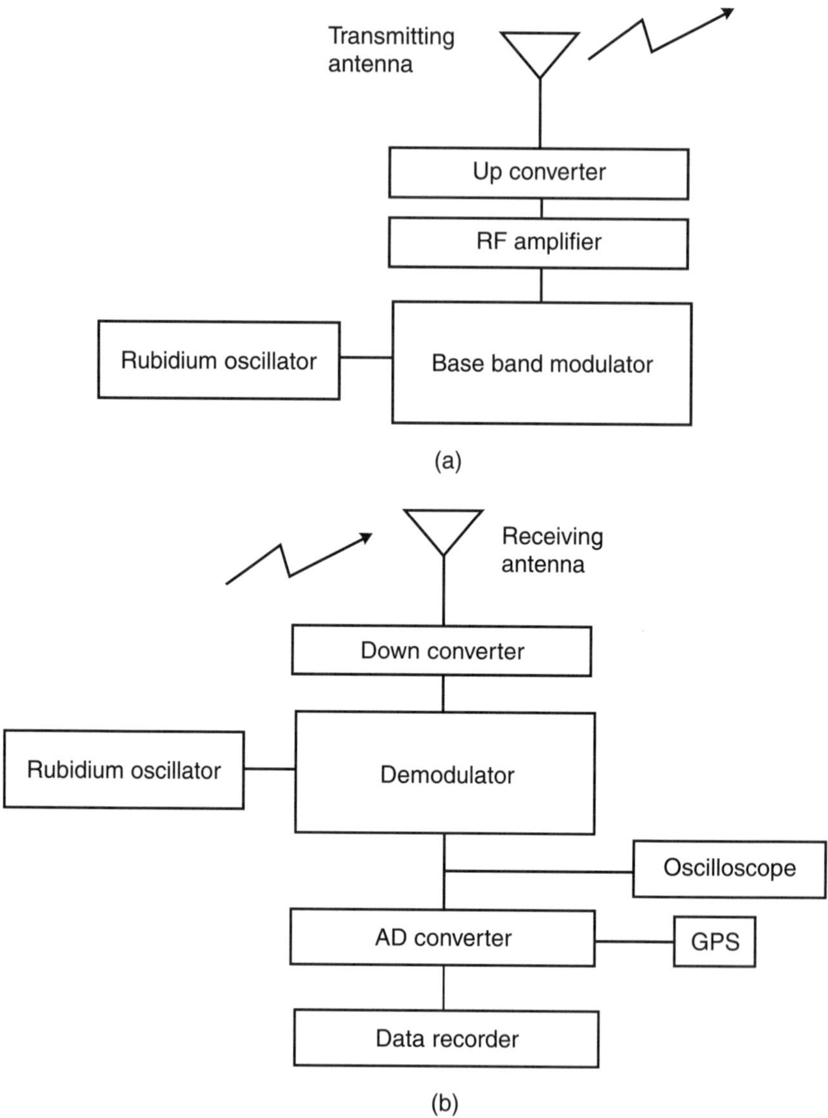

Figure 1.13 Delay profile measurement system: (a) transmitting system and (b) receiving system.

transmitter and the receiver. If the receiving points are located close to the transmitter, a common signal generator source can be used to supply the timing signal to both the transmitter and the receiver, using coaxial cable. For mobile terminal measurements, however, the use of cables is not feasible. A rubidium oscillator in a standard signal generator synchronizes the oscillators used in the transmitter and the receiver.

When performing a delay profile measurement indoors, the transmitting and receiving antennas can both be connected to a vector network analyzer to obtain the received signal time-domain characteristics. After performing the measurement over a wide frequency range, the transmission coefficient (S21) frequency data are transformed from the frequency domain into the time domain using Fourier transformations. The definition of the *S* parameters measured by the network analyzer are described in Chapter 2. This is the simplest method for obtaining the time-domain delay profile. However, note that the measurement resolution is dependent on the frequency bandwidth of the antennas used.

1.5.1.4 Base Station Maintenance and Testing

A propagation measurement is also required when checking the performance of the base station antenna and the corresponding coverage area. This measurement is made using a receiver placed on a vehicle to reduce the time required to perform a measurement over a wide coverage area [8]. Figure 1.14 shows a typical on-board receiver configuration used to measure the received signal strength. The system consists of the following subsystems: a location detector, a data processor, and an electric field strength measuring device. The location detection subsystem is used to estimate the precise position of the vehicle and consists of a directional sensor, a distance sensor, route history data, and a road map stored in a CD-ROM to help generate the measurement route history. Multiple measuring receivers are also placed on board to allow the simultaneous measurement of multiple radio channels.

1.5.2 Calibration

For precise propagation measurements the calibration procedure is very important. For this type of measurement two types of calibration procedure are generally used. The first involves the linearity compensation of the receiver in systems such as those described in Sections 1.5.1.1, 1.5.1.2, and 1.5.1.4. The second type of calibration procedure is that of time standard calibration and is generally used in the measurement system described in Section 1.5.1.3.

The calibration of the RF system is carried out by connecting the signal generator output directly to the antenna input port of the RF receiver. Figure 1.15 shows a calibration chart of the detector system output power versus the detected signal level. This chart can be used to help compensate for the effects of the detector nonlinearity in the measured received signal

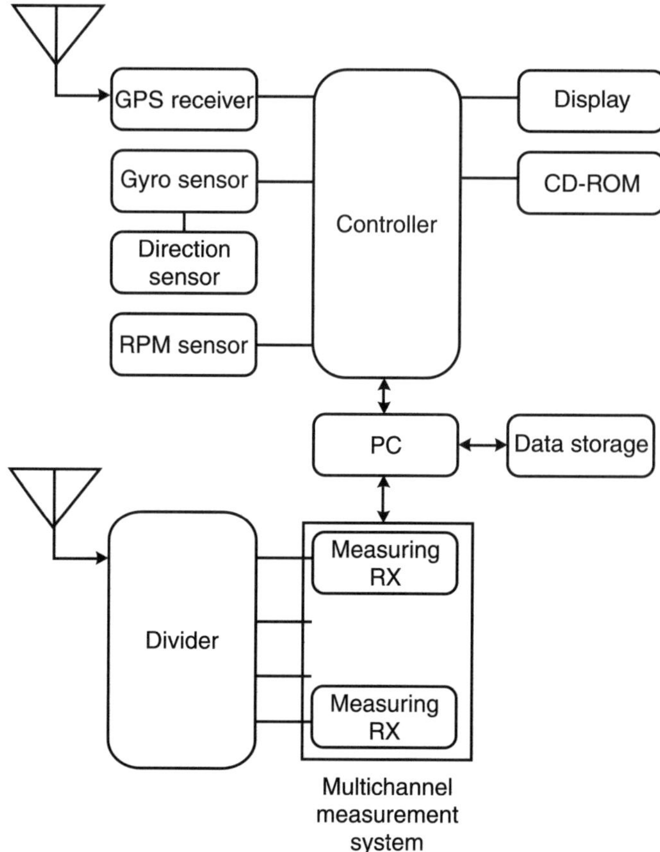

Figure 1.14 Field strength measurement system.

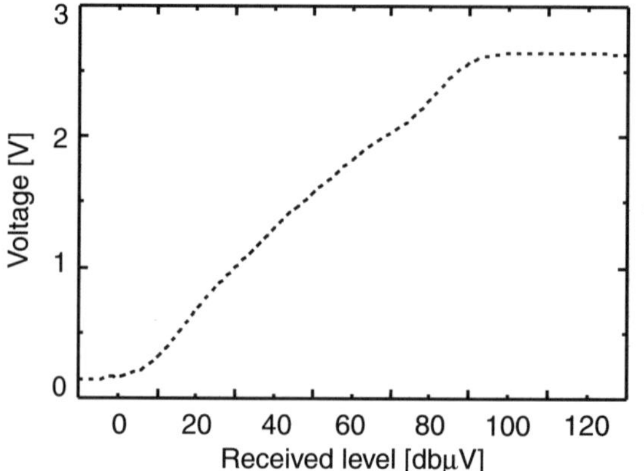

Figure 1.15 Detector calibration curve.

data. This detector system calibration is best carried out in advance of the actual measurement.

To synchronize the phase of the signal between the transmission and reception systems, the two rubidium oscillators are synchronized before the actual measurement. However, this synchronization may last for a period of only a few days and for a precise measurement the synchronization procedure must be repeated every 2 to 3 hours.

1.5.3 Antenna Switching Measurement System

The conventional propagation measurement system uses a diversity antenna in conjunction with the corresponding number of calibrated receivers to detect several input signals simultaneously. As mentioned previously, the received signal envelope is recorded on a laptop computer at regular sampling intervals in time. For indoor propagation measurements, several receivers are required to obtain the characteristics of the different diversity branches. Each time a measurement is made, the system must be calibrated. A simple calibration-free measurement procedure using a single receiver with an antenna switching unit is presented in this section. This system is described in detail because it forms the basis for all propagation measurement systems.

Figure 1.16 shows the block diagram of the measurement system. An antenna switch consisting of three GaAs chips with dummy loads selects the received signals alternatively from two antennas. When the switch selects one antenna, the other antenna is terminated by a 50Ω dummy load. This

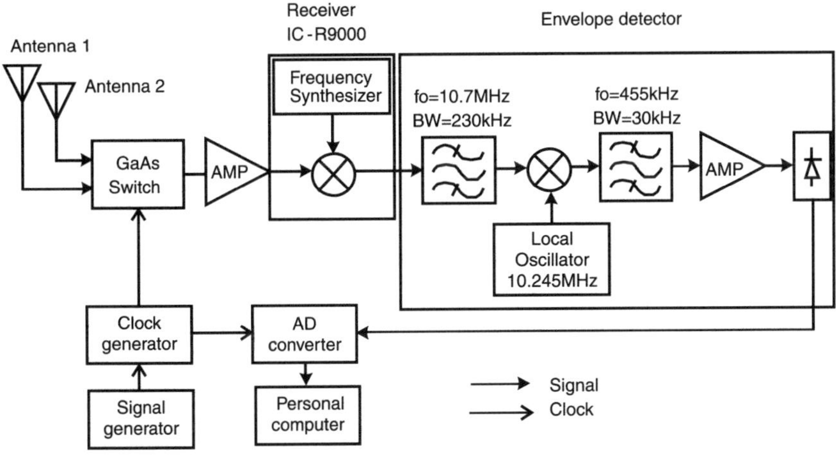

Figure 1.16 Antenna switching measurement system.

provides an isolation of 60 dB and an insertion loss of 1.3 dB below 1 GHz. This type of switch is inexpensive compared with a pin diode switch and is ideal for low-cost measurement systems. The received signal is amplified by 15.1 dB and is then converted to an IF signal (at 10.7 MHz) by a receiver IC chip (IC-R9000). The IF signal is then detected by a standard envelope detector of a type used in FM IF transceivers and portable phones.

To obtain a rapid system response, a wide-bandwidth filter must be used. The bandwidths selected for this particular system were 230 kHz for the first filter and 30 kHz for the second. The detected envelope signal level is recorded in a digital data format using a two-channel analog-to-digital (A/D) converter. The measurement system is compact and can be placed on a portable trolley for convenience.

1.5.4 Antenna Switching Rate

To characterize diversity antennas, several antenna signals are received simultaneously. When the sampling interval is small enough to neglect the time deviation between samples, the signals may be considered to be sampled simultaneously [7]. The maximum sampling rate gives an estimate of the short-range measurement parameters.

Figure 1.17 shows the antenna used for the measurements. This particular space diversity antenna consists of two quarter-wavelength monopoles

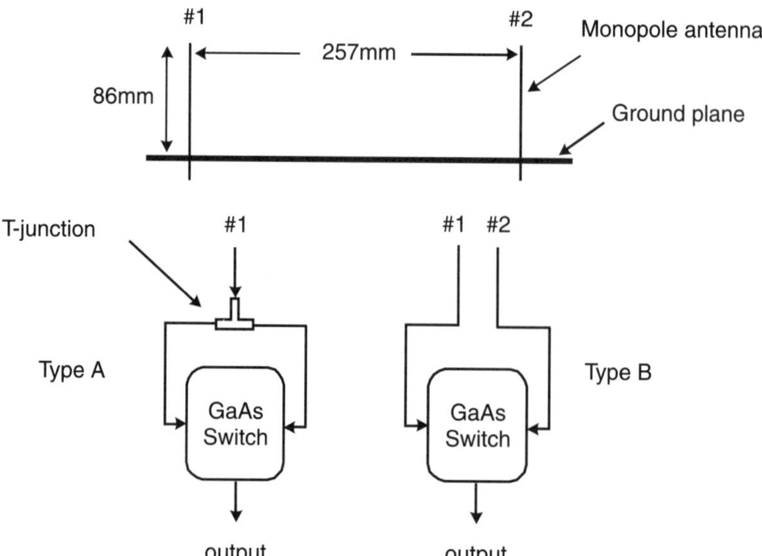

Figure 1.17 Antenna for measurement and feed arrangement for type A and type B circuits.

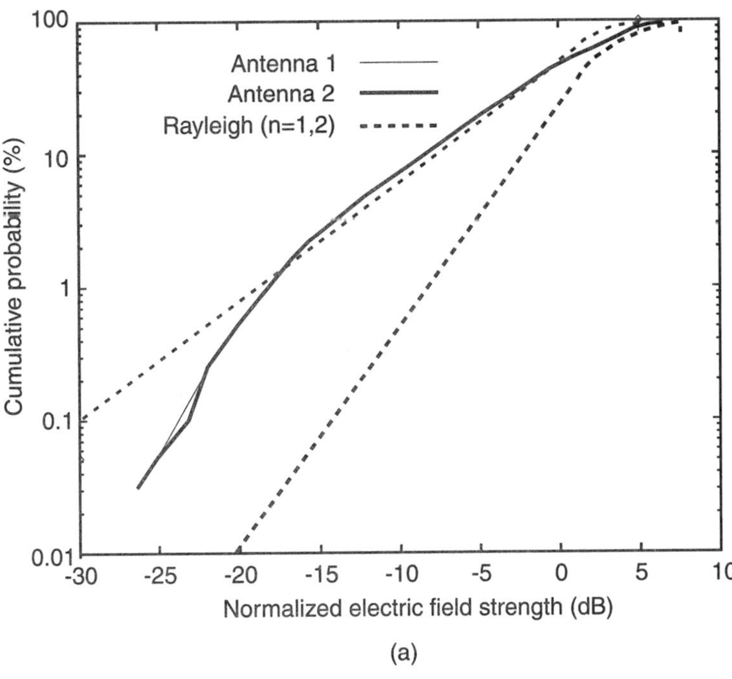

(a)

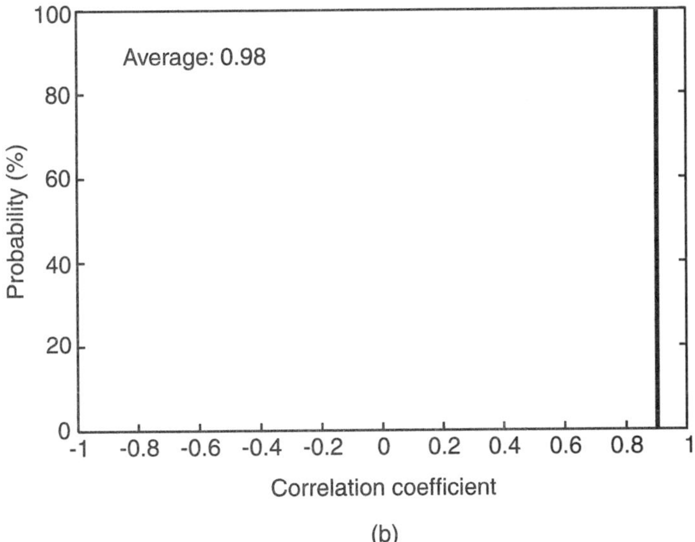

(b)

Figure 1.18 (a) Cumulative probability and (b) the histogram of the correlation coefficient for the type A diversity antenna.

with three-quarter-wavelength spacing. Two feeder circuits are used for this antenna. For type A, a T junction separates the monopole antenna output into two ports, and the other antenna is not used. For type B, the two antenna outputs are connected to the antenna switch. When the switching rate

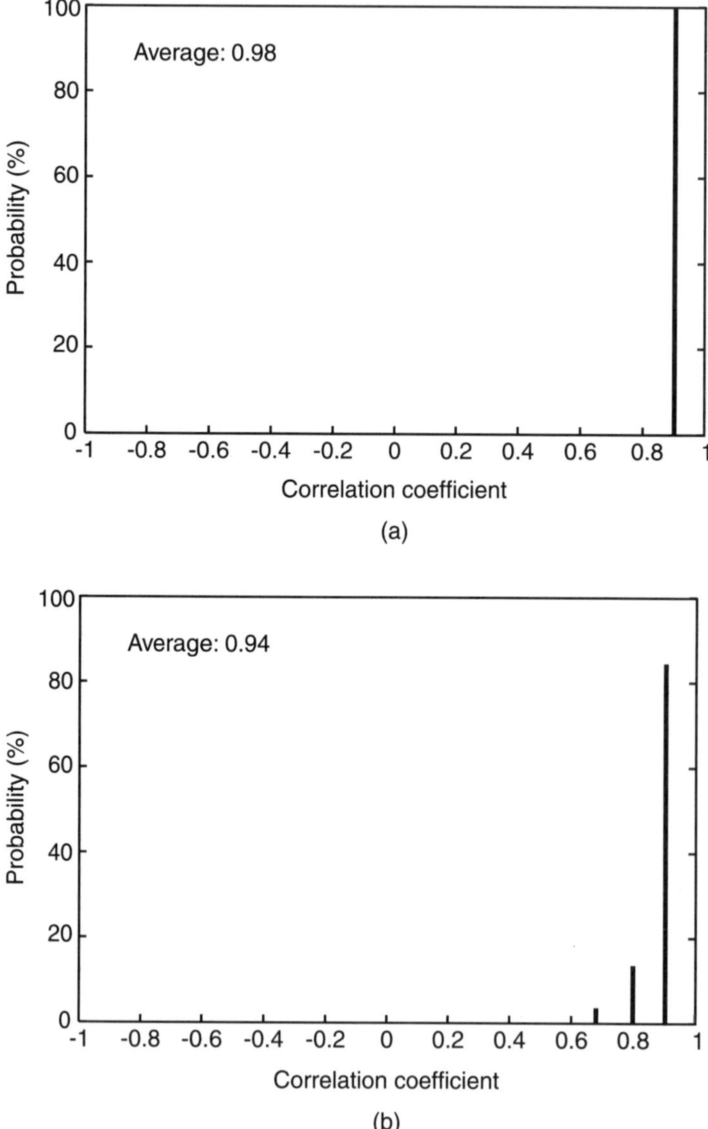

Figure 1.19 Histograms of the correlation coefficient (a) for a switching interval of λ/40 and (b) for a switching interval of λ/20.

is fast enough to neglect the time deviation between adjacent samples, the correlation is 1.0 for type A, and 0.1 for type B.

The antennas are moved along an indoor passageway from end to end at a uniform velocity so that in one switching cycle they move a distance equal to $\lambda/200$. The corresponding switching frequency is 330 Hz and the total number of data samples n is 25,000.

The two curves in Figure 1.18 are almost identical and follow Rayleigh distribution above 3% to 4%. The histogram of the correlation is entirely contained within the interval 0.9 to 1.0 with a mean value of 0.98.

If it is possible to take a large sampling interval in the measurements, the antenna switching system provides a multibranch diversity measurement. The maximum achievable sampling interval can be simulated using data samples from the type A antenna measurement. Figure 1.19 shows the histogram of correlation for sampling intervals of $\lambda/40$ and $\lambda/20$. These results are obtained by thinning the data in Figure 1.18. The histogram in Figure 1.19(a) is identical to that of Figure 1.18, but that of Figure 1.19(b) differs substantially. These results indicate that an interval of $\lambda/40$ is sufficiently short that the received signal envelope can be considered constant between measurements. The corresponding switching rate of 66 Hz is short enough to apply to a stationary environment, because the velocity of people moving in the building is less than 1 m/s.

For this particular diversity antenna system, with a sampling interval of $\lambda/40$ and the antenna switching rate indicated above means that the sampling rate for a single antenna is $\lambda/20$. This is the reason for choosing the envelope signal level as described in Section 1.5.1.1 for short-range measurements without the use of interpolation.

References

[1] Jakes, W. C., Jr., *Microwave Mobile Communications,* New York: John Wiley & Sons, 1974.

[2] Clarke, R. H., "A Statistical Theory of Mobile-Radio Reception," *Bell Syst. Tech. J.*, Vol. 47, No. 6, July–Aug. 1968, pp. 957–1000.

[3] Hata, M., "Empirical Formula for Propagation Loss in Land Mobile Radio Services," *IEEE Trans. Vehicular Technology*, Vol. VT-29, No. 3, 1980, pp. 317–325.

[4] Taga, T., "Analysis for Mean Effective Gain of Mobile Antennas in Land Mobile Radio Environments," *IEEE Trans. Vehicular Technology*, Vol. VT-39, No. 2, May 1990, pp. 117–131.

[5] Lee, W. C. Y., *Mobile Communication Engineering,* New York: John Wiley & Sons, 1982.

[6] Nakano, M., et al., "Up-Link Polarization Diversity Measurement for Cellular Communication Systems Using Hand-Held Terminal," *Conf. Digest IEEE AP-S Int. Symp.*, Montreal, Canada, July 13–18, 1997, pp. 1360–1363.

[7] Murase, M., K. Tanaka, and H. Arai, "Propagation and Antenna Measurements Using Antenna Switching and Random Field Measurements," *IEEE Trans. Vehicular Technology*, Vol. 43, No. 3, 1994, pp. 537–541.

[8] Komizu, T., et al., "Development and Field Test Results of a Digital Cellular System in Japan," *Proc. IEEE Vehicular Technology Conf.*, Stockholm, Sweden, 1994, pp. 302–305.

2

Antenna Measurements for Radio Handsets and Mobile Terminals

In the design and development of new antennas for radio handsets and mobile terminals, a prime consideration is the impedance matching requirements of the antenna input port. Once the antenna input port impedance has been suitably matched, the next parameter to be measured is the antenna radiation pattern. The performance of a newly designed antenna can be measured in an indoor location before final testing in a real propagation environment.

The first sections of this chapter deal with parameter definitions for impedance and radiation pattern measurements. To produce miniature communications products, electrically small antennas are often used. However, electrically small antennas have the disadvantages of a narrow-frequency bandwidth and low radiation efficiency. A typical example of this type of antenna is a pager antenna.

The frequency bandwidth characteristics of an antenna can be obtained by measuring the antenna impedance over a range of frequencies. To obtain the antenna radiation efficiency, another measurement approach is required that is dealt with in the later sections of the chapter. Some mobile communications systems require the incorporation of diversity function antennas. As was described in Chapter 1, the diversity performance of the antenna is usually obtained by propagation measurements. However, the diversity characteristics may also be obtained by measuring the complex radiation pattern of the antenna and by utilizing a mathematical model that describes the arrival of electromagnetic waves from various regions of the space surrounding the antenna. The third part of the chapter describes this method.

In addition to the above measurements, an electromagnetic compatibility (EMC) or electromagnetic emission (EMI) test is a mandatory requirement before releasing the antenna as a commercial product. This is done to ensure that the antenna meets national wireless regulatory standards. A typical measurement method used for this purpose is described in the latter parts of the chapter. In the final section of the chapter, mobile terminal antenna tests used for product evaluation are presented. This chapter deals with the necessary measurement procedures for handset and mobile terminal antennas. The measurement procedures required for the evaluation of the base station antenna performance are dealt with in Chapter 4.

2.1 Antenna Input Port Impedance Measurements

Most radio-frequency (RF) output ports are designed to have an impedance of 50Ω, thus the antenna input impedance is also required to be matched to this value. The impedance characteristics of the antenna can be found by measuring the voltage standing-wave ratio (VSWR), otherwise called the return loss. However, before giving details of this measurement procedure, the definitions of some basic terminology are required.

2.1.1 Reflection Coefficient

A definition for the reflection coefficient can be derived from the model shown in Figure 2.1, which illustrates an impedance mismatch situation be-

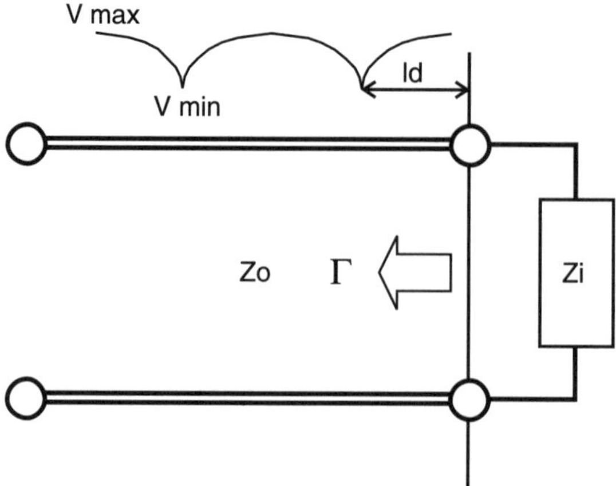

Figure 2.1 Reflection coefficient from antenna.

tween a transmission line of impedance Z_o and a terminating load of impedance Z_i (the antenna). The reflection coefficient Γ for the terminating impedance is defined according to the following equation:

$$\Gamma = \frac{Z_i - Z_o}{Z_i + Z_o} \tag{2.1}$$

In actual measurements, the reflected power at the antenna input port is used as a measure of the impedance mismatch factor. This reflected power is referred to as the return loss (RL). In the field of antenna engineering, the RL is defined as negative, although, in general, in other fields, RL is defined as positive.

$$\text{RL} = 20 \ \log_{10} |\Gamma| \tag{2.2}$$

As shown in Figure 2.1, the impedance mismatch between the transmission line and the antenna sets up a standing wave along the transmission line. The ratio between the maximum voltage $V_{\max}$ and the minimum voltage $V_{\min}$ then defines the VSWR, denoted by ρ, as follows:

$$\rho = \frac{V_{\max}}{V_{\min}} = \frac{1 + |\Gamma|}{1 - |\Gamma|} \tag{2.3}$$

Thus, the degree of impedance mismatch between the transmission line and the antenna can be defined by the VSWR. Examples of the maximum impedance mismatch value to be expected for standard communication system antennas are $\rho \leq 1.1$ for high-power broadcasting station transmitting antennas, $\rho \leq 1.5$ for cellular base station antennas, and $\rho \leq 2$ or $\rho \leq 3$ for low-gain mobile terminal antennas.

2.1.2 Voltage Standing-Wave Ratio, *S* Parameters, and the Smith Chart

The classical approach chosen for measuring the VSWR utilizes coaxial waveguides, as shown in Figure 2.2. The standing wave excited inside the coaxial waveguide is measured directly by a probe inserted through a slit made in the waveguide's outer conducting wall. A standard measurement procedure that utilizes this method is described in this section. Initially, the physical

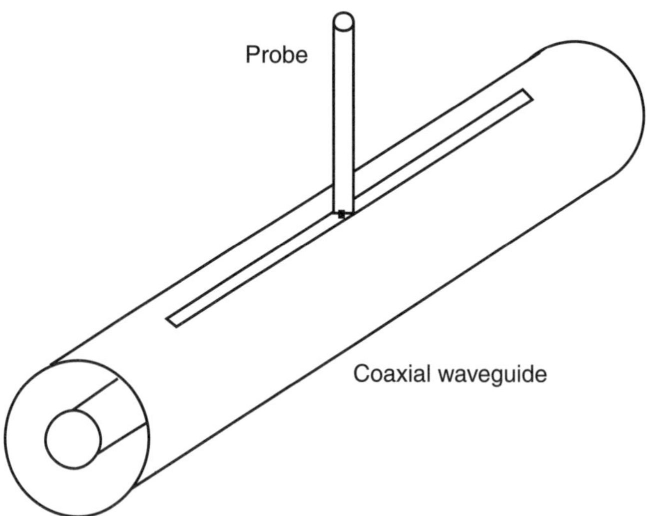

Figure 2.2 VSWR measurement by coaxial waveguide.

location of the first node of the standing wave for the coaxial cable under short circuit conditions is marked. As can be seen in Figure 2.1, when the device under test (DUT), in this case an antenna, is connected to the end of the coaxial waveguide, the node position is found to shift toward the DUT terminal point, in this case to a distance l_d from the DUT terminal.

The impedance of any device may be illustrated in a 2-D complex plane representation referred to as the Smith chart (Figure 2.3) [1]. This chart represents a mapping of constant-resistance and constant-reactance components of the DUT onto the complex reflection coefficient plane, using the following equation:

$$\frac{Z_l}{Z_o} = \frac{1+\Gamma}{1-\Gamma}, \qquad \hat{z} = \frac{Z_l}{Z_o} = z_r + jz_x \tag{2.4}$$

where $\hat{z}$ is the normalized load impedance. Setting z_r and z_x equal to any real constants yields two transformations from coordinate straight lines in the complex $\hat{z}$ plane to circles in the complex $\Gamma = \Gamma_r + j\Gamma_i$ plane. Any z_r and z_x constant lines are given by the following circles:

$$\left(\Gamma_r - \frac{z_r}{z_r+1}\right)^2 + \Gamma_i^2 = \left(\frac{1}{z_r+1}\right)^2 \tag{2.5}$$

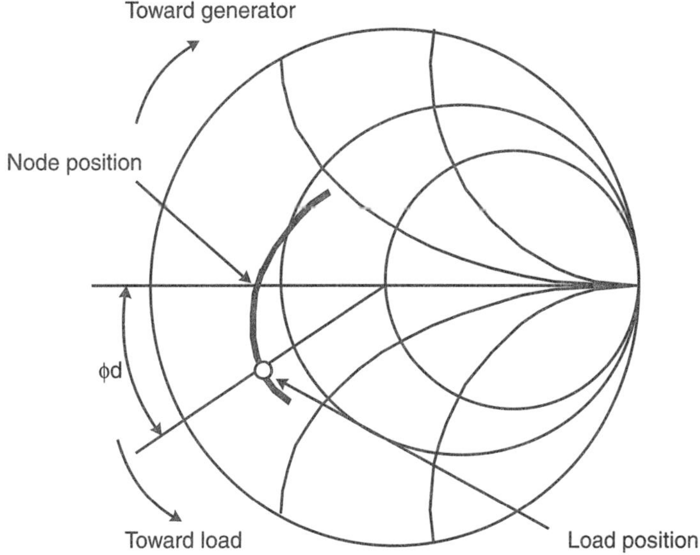

Figure 2.3 Impedance measurement by Smith chart.

$$(\Gamma_r - 1)^2 + \left(\Gamma_i - \frac{1}{z_x}\right)^2 = \left(\frac{1}{z_x}\right)^2 \tag{2.6}$$

A circle of radius ρ is drawn on the Smith chart to represent the VSWR value. The normalized load impedance can be obtained from the Smith chart by rotating the measuring point from the initial node position for the waveguide under short circuit conditions by an angle equivalent to $4\pi l_d / \lambda_g$ toward the load side of the chart.

This fundamental measurement method is also used for impedance mismatch measurements in high-power transmission lines. The most commonly used device for measuring the impedance of RF devices is a network analyzer. The input impedance of an individual dipole, as well as the mutual impedance of a dipole pair (Figure 2.4), can be measured using a network analyzer. The terms a_1, a_2 and b_1, b_2 denote the amplitude of the incident and reflected waves from dipole antennas 1 and 2, respectively. The reflection and transmission coefficients S_{ij} are expressed in the form of the scattering matrix:

$$\begin{pmatrix} b_1 \\ b_2 \end{pmatrix} = \begin{pmatrix} S_{11}\, S_{12} \\ S_{21}\, S_{22} \end{pmatrix} \begin{pmatrix} a_1 \\ a_2 \end{pmatrix} \tag{2.7}$$

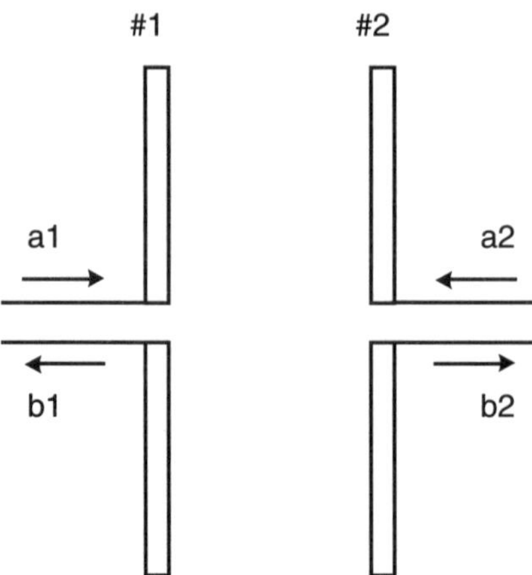

Figure 2.4 *S* parameter measurement of a pair of dipoles.

The return loss (RL) and isolation factor (IS), which are often used as antenna measurement parameters, can be defined as follows:

$$RL = 20 \log_{10} |S_{11}|, \qquad IS = -20 \log_{10} |S_{12}| \tag{2.8}$$

It is the *S* parameters shown in the scattering matrix that are directly measurable by the network analyzer. As can be seen from Figure 2.5, the network analyzer has two measurement ports. The complex signal levels extracted via the directional coupler (DC) from the DUT are denoted by C_1 through C_4 and can be expressed in the form of the scattering matrix:

$$\begin{pmatrix} S_{11} & S_{12} \\ S_{21} & S_{22} \end{pmatrix} = \begin{pmatrix} \dfrac{C_2}{C_1} & \dfrac{C_2}{C_4} \\ \dfrac{C_3}{C_1} & \dfrac{C_2}{C_4} \end{pmatrix} \tag{2.9}$$

The impedance matrix, normalized by the characteristic impedance of the transmission line, is evaluated by the following matrix conversion:

$$[z] = \{[1] - [S]\}^{-1}\{[1] + [S]\} \tag{2.10}$$

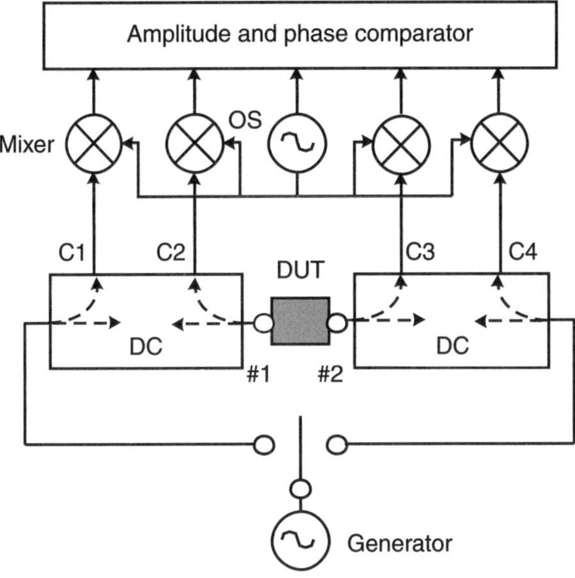

Figure 2.5 Diagram of network analyzer.

where $[z]$, $[S]$, and $[1]$ represent the normalized impedance matrix, the scattering matrix, and the unit matrix, respectively. Thus, the input and mutual impedances are given by the measured value obtained directly from the network analyzer.

Other important points that affect the correct usage of the network analyzer include sufficient time to allow the instrument to warm up before performing the measurement, correct instrument calibration, and a proper reference plane for the connectors. For stable network analyzer operation, the instrument must be allowed to warm up for a period of 30 to 60 minutes. The calibration procedure to obtain correctly referenced amplitude and phase values of the signals under analysis must be carried out after the instrument warming-up period is complete using a calibration kit supplied by the instrument manufacturer. To obtain measurements of as high a precision as possible, the minimum frequency bandwidth used during measurements must be covered by the calibration process. Finally, to derive the correct impedance value, the position of the reference plane of the measurements must also be taken into account. As can be seen from Figure 2.6, the reference plane for the measurements is normally located on the surface of the dielectric material on the inside of the connector.

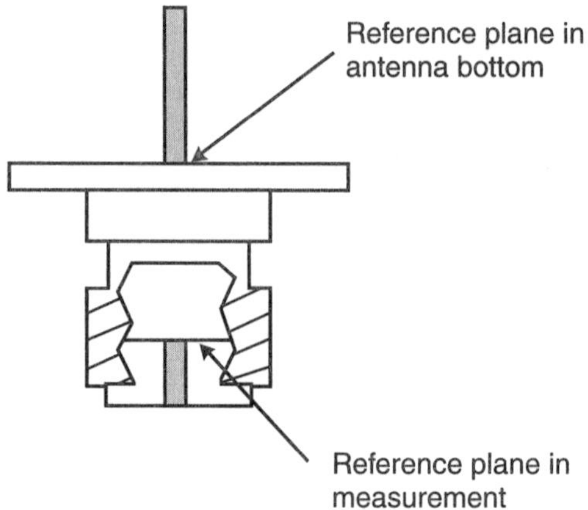

Figure 2.6 Reference plane of connector.

2.1.3 Balanced and Unbalanced Feeder Systems

Feeder transmission lines for antennas can be classified into two kinds: balanced or unbalanced. The Lecher line (shown in Figure 2.7) is an antenna feeder composed of a pair of parallel conducting wires that supplies an out-of-phase current and voltage. However, in current mobile communication systems, it is more common to use a coaxial cable feeder. The coaxial cable is classified as an unbalanced line because the voltage on the outer conductor of the cable is always at zero potential.

Another name for this type of transmission cable is a *shielded cable.* It is referred to as a shielded cable because the outer conductor is connected to electrical earth and effectively acts to shield the cable from the effects of external RF noise. These two different types of transmission line cannot be

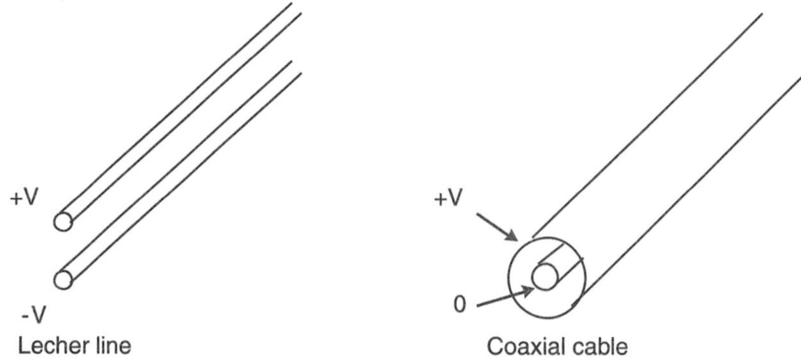

Figure 2.7 Voltage of Lecher line and coaxial cable.

connected together directly but must be connected via a voltage transformer (as shown in Figure 2.8). The voltage transformer consists of a pair of inductors and a pair of capacitors connected in the form of a bridge circuit and is effective in the VHF frequency region. A transformer called a *balun* is required for the UHF frequency band.

In antenna measurements, the half-wavelength dipole is often used as a standard antenna. This antenna consists of two electrical poles excited by a balanced transmission line. For measurement purposes, the coaxial cable is an appropriate transmission line because the intrinsic shielding ability of the cable ensures that there is no RF interference with surrounding equipment. If, however, a dipole antenna is fed by a coaxial cable, undesired currents will flow on the outer conductor of the cable. This leakage current can be suppressed by using a balun-type structure as shown in Figure 2.9. The *bazooka balun* consists of a double coaxial cable. Because the balun has an electrical length of one-quarter, the impedance at the mouth of the cable becomes infinite, effectively suppressing the leakage current from the dipole.

In addition to the balun, in the UHF band a ferrite ring can also serve as an impedance transformer and this is often used in mobile handset receiving antennas. To suppress the leakage current effectively, the coaxial cable can be wrapped around a ferrite bead, but this process incurs additional losses in the antenna. In this case, the leakage current is dissipated as thermal losses. In handset antenna designs, the chassis of the handset is modeled as a conducting box, as shown in Figure 2.10. A monopole antenna creates a leakage current on the chassis because the box acts as the complementary virtual pole of the monopole.

The antenna leakage current flows on the outside of the handset chassis and has a great influence on the resulting antenna radiation pattern. Mobile phone handsets also employ several built-in antennas. However, leakage current to the handset chassis also causes similar problems in the antenna

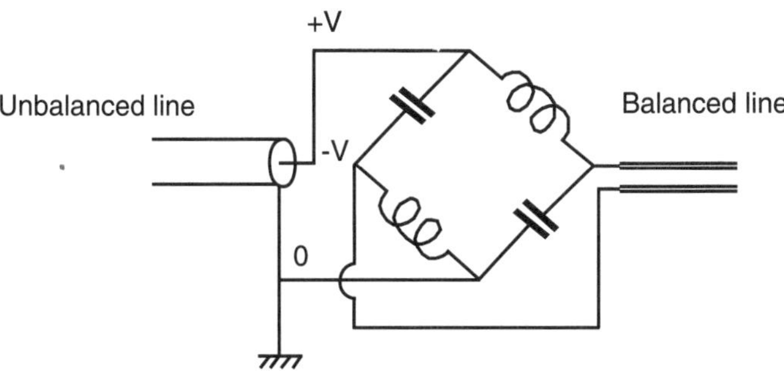

Figure 2.8 Transformer circuit for balanced and unbalanced line.

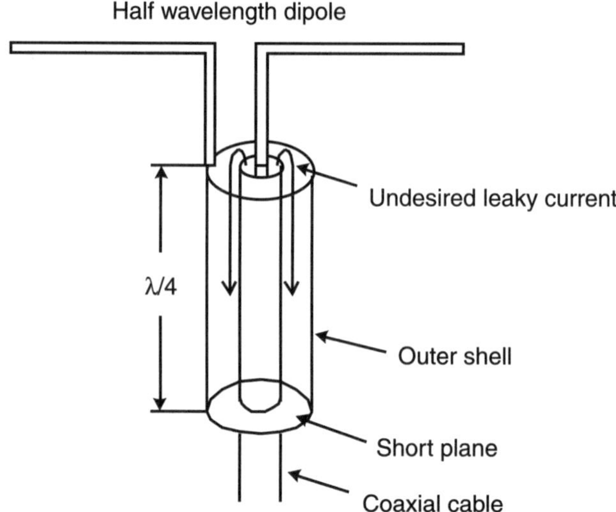

Figure 2.9 Dipole antenna fed by coaxial cable using balun structure.

radiation characteristics. This is also true when performing antenna measurements on handset antennas. The leakage current flowing over the antenna chassis leaks to the coaxial cable and changes the antenna characteristics. Thus, in order to suppress this leakage current, a balun is necessary for performing proper antenna impedance measurements. Several basic types of balun are discussed in [2].

Dipole antennas fed by a balanced line are sometimes used in mobile communication systems. When it is necessary to measure the input imped-

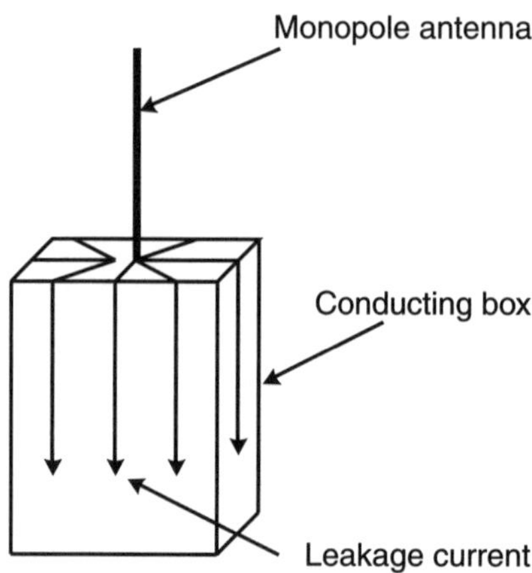

Figure 2.10 Model of handset antenna and leakage current on the box.

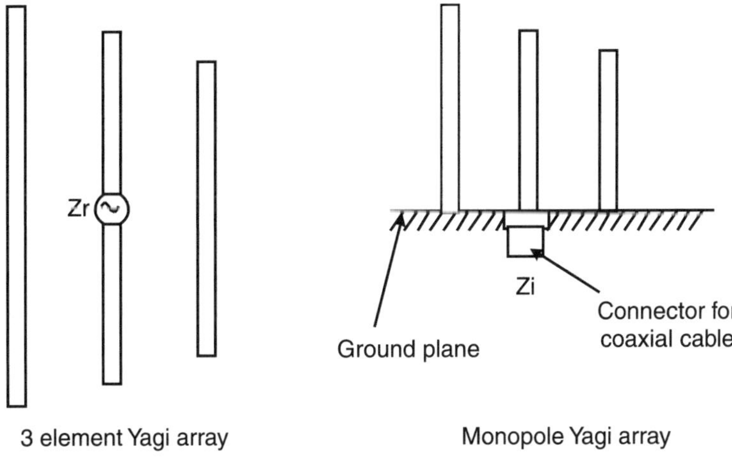

Figure 2.11 Impedance measurement of Yagi array using image method.

ance characteristics of the balanced line feeding the antenna, a method referred to as the image method is used that utilizes a network analyzer. Figure 2.11 shows a typical measurement setup for a Yagi array antenna. The antenna structure is divided into two halves about the central symmetrical axis. One-half of the antenna is mounted on a large electrical ground plane including the antenna feed point. This method is used to measure the input impedance of the antenna. Physically, the measured antenna is only one-half of the actual antenna structure, and thus the true input impedance of the antenna is twice the measured impedance value.

2.1.4 Effect of Finite Size Ground Plane

Compact planar antennas used in mobile communications applications are generally mounted on an electrical ground plane and fed via a coaxial cable. The impedance characteristics of the antenna are heavily dependent on the area of the conducting ground plane. If the ground plane does not have a sufficient area, the measured impedance characteristics of the antenna become a function of the area and also prove to be different from those predicted for an infinite ground plane.

In practical applications, it is impossible to mount the planar antenna on an infinite ground plane; however, to obtain good impedance characteristics, a circular ground plane with a radius of 1.5λ can be used as the required minimum area. This value is chosen because the impedance then converges to within 5% of the infinite plane value, as can be seen in the simulation shown in Figure 2.12. The calculations have been performed using the

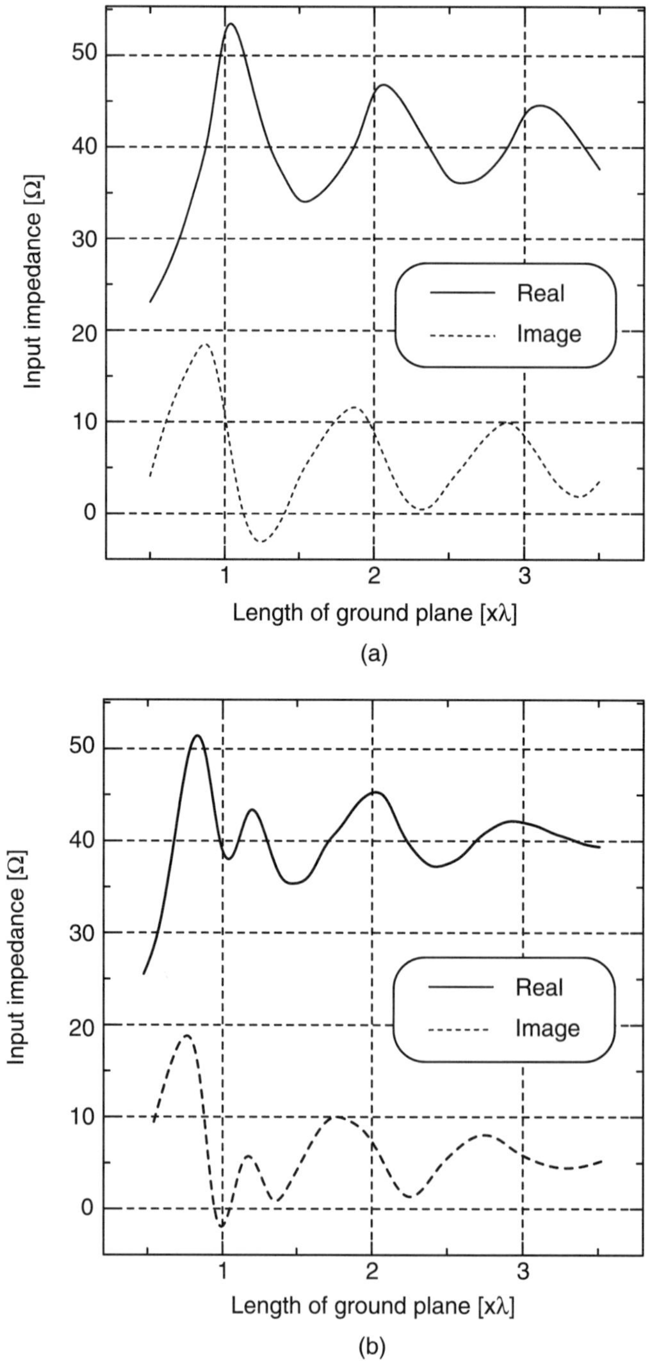

Figure 2.12 Input impedance of monopole antenna on finite size ground plane: (a) disk ground plane and (b) square ground plane.

finite difference time-domain (FDTD) method [3], and give the same results as the method of moments (MOM) [4]. The numerical simulation shows that the input impedance is very unstable for ground plane lengths of less than 1.3λ. If the ground plane is too small, the current spreads to the edges of the plane and flows along the rear surface and along the outer conductor of the feeder. A leakage current flowing on the outside of the conducting coaxial feeder causes this instability. If the ground plane is of sufficient size, this current decreases to zero before reaching the edges of the plane and thus gives a more stable impedance characteristic.

Another possible shape of finite ground plane is square, and Figure 2.12(b) shows the input impedance of a monopole antenna mounted on a square-shaped ground plane as a function of the side length. The input impedance convergence is better for a square-shaped ground plane than for a circular one. The reason for this is that the current return path for the case of the circular ground plane is of the same length for all return paths, while most of the current paths are different for the square ground plane. However, it has the disadvantage that it creates a large cross-polarization component in the horizontal plane caused by the different length current paths. Figure 2.13 shows the radiation patterns in the horizontal plane for circular and square-shaped ground planes. The cross-polarization level is suppressed well by the circular ground plane and this is an advantage of the use of this shape.

2.2 Radiation Pattern Measurements

Another important antenna design parameter is the far-field radiation pattern. The radiation field for an antenna consists of a static field region (near field) close to the antenna and a radiative field region (far field) further away from the antenna. In the case of the radiative field region, the electric field strength is, in general, inversely proportional to the distance from the antenna. To understand the antenna radiation field, the near- and far-field distances must be clearly defined. This section presents the parameters that define the far-field distance from the antenna and also describes a measurement system for the far-field radiation pattern. A near-field measurement system is described in Chapter 4.

2.2.1 Definition of Far Field

Calculating the phase error due to the finite size of the antenna defines the far field. Figure 2.14 shows the difference in the path length from the

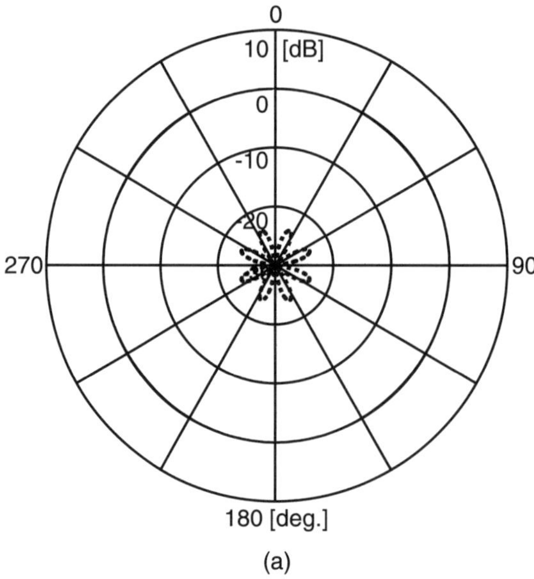

(a)

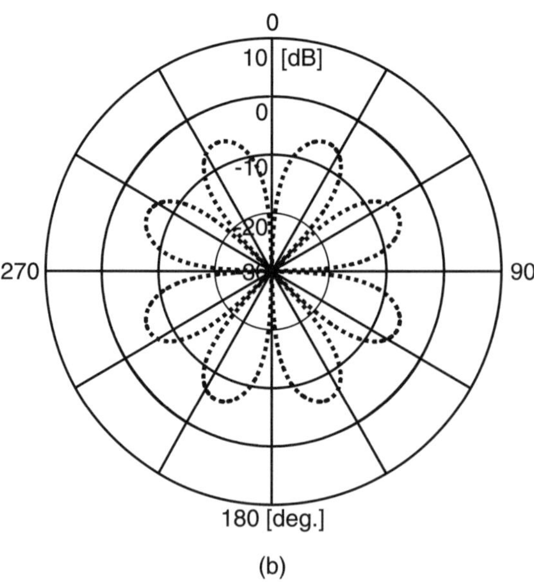

(b)

Figure 2.13 Cross-polarization radiation pattern in horizontal plane of monopole antenna on finite size ground plane: (a) disk ground plane with diameter of 1.5λ and (b) square ground plane with side length of 1.5λ.

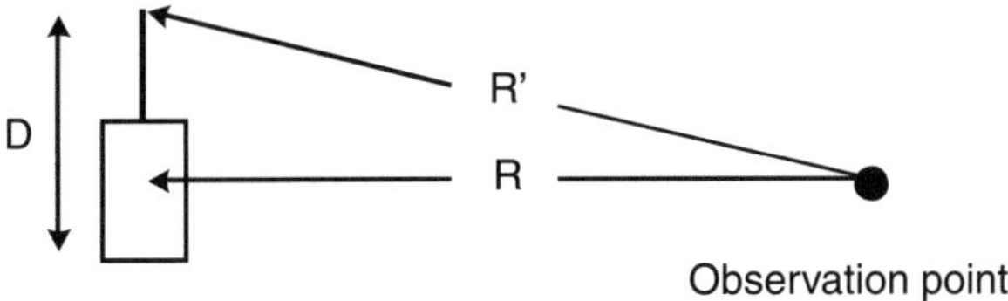

Figure 2.14 Definition of the far field.

observation point to the base of the antenna and to the end of the antenna. This path length difference can be translated into a phase difference as indicated by the equation:

$$\Delta\phi = \frac{2\pi}{\lambda}(R' - R) = \frac{2\pi}{\lambda}\left\{\sqrt{R^2 + \left(\frac{D}{2}\right)^2} - R\right\}$$

$$\cong \frac{2\pi}{\lambda}\frac{1}{R}\left(\frac{D}{2}\right)^2 = \frac{\pi D^2}{4\lambda R} \tag{2.11}$$

For the phase error to be less than $\pi/8$, the distance R from the antenna must be such that:

$$R \geq \frac{2D^2}{\lambda} \tag{2.12}$$

The above equation is often used to define the far-field measurement distance for the antenna. This phase error, however, limits the dynamic range of the radiation pattern. If the distance R satisfies the condition of (2.12), a measured pattern of less than –23 dB of the peak radiation level contains measurement error [5].

Thus, for this particular case, the dynamic range for a phase error of $\pi/8$ is 23 dB. It is therefore meaningless to discuss radiation field levels below this 23-dB level for a phase error definition of $\pi/8$. If, however, the defined phase error level is halved to $\pi/16$, then the far-field distance, as defined by (2.12), is doubled and the dynamic range of the antenna will also be increased to 25 dB. Thus, when defining the far-field distance of the antenna, the relationship between the phase error and the dynamic range must be kept in mind.

2.2.2 Measurement System Setup

We now discuss the setup of the measurement system [6]. In far-field measurements the distance between the transmitter and the receiver should be sufficiently large to satisfy the far-field condition stated earlier. For large base station antennas with high gain in mobile communication systems, the antenna radiation field is measured in an outdoor environment. The measurement environment should be in a location where there is very little RF noise present. The specifics of site selection as well as details concerning the measurement apparatus required are described in Chapter 4. For the measurement of low- and medium-gain antennas used on mobile terminals, an indoor measurement environment may be used. Details concerning the chamber geometry and the measurement equipment requirements for these types of antennas are presented in this section.

The most common type of RF anechoic chamber has a square cross-section with electromagnetic wave absorber placed on the inside walls of the chamber, as shown in Figure 2.15(a). The transmitting antenna and the antenna device under test (DUT) are placed near the side walls of the chamber to satisfy a reflection wave level of less than –30 dB. This region is referred to as the "quiet zone" and is defined as a sphere of given radius centered on the antenna concerned. The measured antenna should be inside this sphere.

In the UHF band, the type of wave absorber used consists of urethane foam in which carbon particles have been embedded. For normally incident waves, the level of signal reflection is reduced to less than –20 dB. For the case of obliquely incident waves, the absorber performance is degraded.

To suppress the signal reflection for obliquely incident waves, the RF absorber is shaped in the form of a pyramid. The obliquely incident wave undergoes multiple reflections as it bounces between opposing faces of the pyramidal blocks. With each reflection the wave signal strength is gradually reduced, and the shape of the absorber helps to reduce the level of signal reflected back toward the antenna.

An example of another type of chamber geometry is shown in Figure 2.15(b). This is referred to as a taper-shaped chamber and this type of chamber geometry helps to reduce the overall volume occupied by the chamber. The transmitting antenna is installed at the top of the chamber taper and radiates a quasi-plane wave toward the antenna under test. The reason why a quasi-plane wave is radiated from the transmitting antenna has to do with the proximity of the RF absorber to the antenna. This type of testing chamber can be used for measurements below 1 GHz and is generally of low cost.

An example of the measurement setup is shown in Figure 2.16. The radiating antenna is placed at the center of the quiet zone, and the antenna

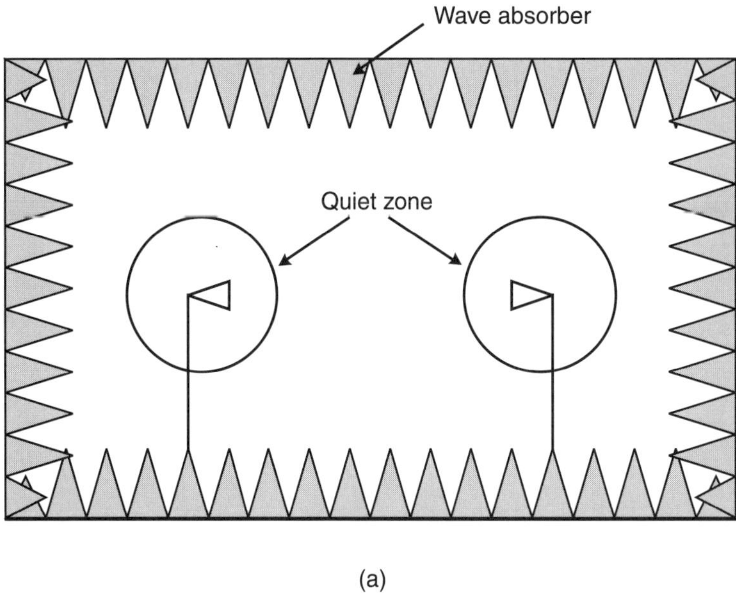

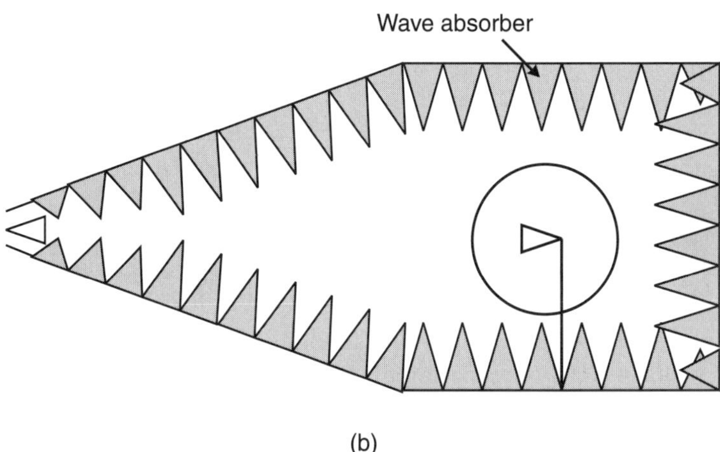

Figure 2.15 (a) Anechoic chamber with a shielded square pillar and (b) taper-shaped anechoic chamber.

under test is mounted on a rotating table on the opposite side of the chamber. The DUT can either be rotated continuously or, using a stepping motor, by fixed angular increments. The signal generator is used to excite the source antenna while the DUT is connected to a receiver. Good system measurement

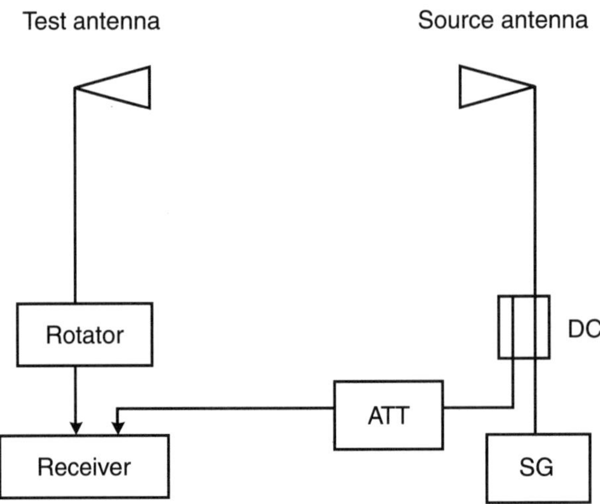

Figure 2.16 Measurement equipment setup.

requires a dynamic range of more than 40 dB and this can be achieved using a phase lock loop (PLL) receiver. At high frequencies (>1 GHz), a harmonic mixer is often placed near the test antenna to decrease transmission losses through the coaxial cable.

To make a PLL receiver system, it is necessary to place a directional coupler in the transmit signal line and to feed a percentage of the transmit signal to the receiver. A personal computer (PC) can be used to control the data acquisition system and the rotation angle of the DUT automatically.

The types of source antenna that are generally used in this particular measurement system include a dipole antenna, a Yagi array, a pyramidal horn, a double-ridged horn, and a log-periodic antenna. The frequency bandwidth characteristics and cross-polarization level of the source antenna should be taken into account when determining the specifications for the measurement system. For example, the following frequency ranges for standard antennas can be used: dipole, <1 GHz; double-ridged horn or log-periodic, 1 to 20 GHz; pyramidal horn, >10 GHz. These are all examples of antennas with linearly polarized radiation patterns. If a circularly polarized radiation field is required, a helical or loop antenna can be employed.

For measurements on low-gain antennas for which the distance between the antennas is relatively short, a network analyzer can be used. The two antennas are connected to the two ports of the network analyzer and an S21 parameter measurement is performed. This is a comparatively low-cost system but the dynamic range is not very large.

2.2.3 Far-Field Measurements

2.2.3.1 Amplitude Pattern Measurement

For the case of field pattern measurements on linearly polarized antennas, the transmitting and receiving antennas are first aligned so that the polarization vectors are in the same plane. After this is done, the measurement is performed as described in the previous section. On the other hand, for circularly polarized antennas, two measurement procedures are required. Before proceeding to a description of the circular polarization measurement, it is first necessary to define the cross-polarization ratio (XPR) and the axial ratio (AR).

A circularly polarized radiation field is expressed by:

$$\mathbf{E} = E_\theta \mathbf{e}_\theta + jE_\phi \mathbf{e}_\phi \tag{2.13}$$

where E_θ and E_ϕ are the complex electric field amplitudes of the electromagnetic wave in the θ and ϕ directions, respectively, and $\mathbf{e}_\theta$, $\mathbf{e}_\phi$ are the unit vectors in the θ and ϕ directions. If $E_\theta = E_\phi$, then the field is called a circularly polarized field. If, however, E_θ is not equal to E_ϕ, then it is called an elliptically polarized field. Equation (2.13) can also be decomposed into right-handed ($\mathbf{E}_R$) and left-handed ($\mathbf{E}_L$) circularly polarized components, where:

$$\mathbf{E}_R = E_R(\mathbf{e}_\theta - j\mathbf{e}_\phi), \quad \mathbf{E}_L = E_L(\mathbf{e}_\theta + j\mathbf{e}_\phi) \tag{2.14}$$

Figure 2.17 illustrates how the field of an elliptically polarized wave may be split into separate $|\mathbf{E}_R - \mathbf{E}_L|$ and $|\mathbf{E}_R + \mathbf{E}_L|$ electric field components. The amplitude coefficients E_R and E_L are given by:

$$E_R = \tfrac{1}{2}(E_\theta - E_\phi), \quad E_L = \tfrac{1}{2}(E_\theta + E_\phi) \tag{2.15}$$

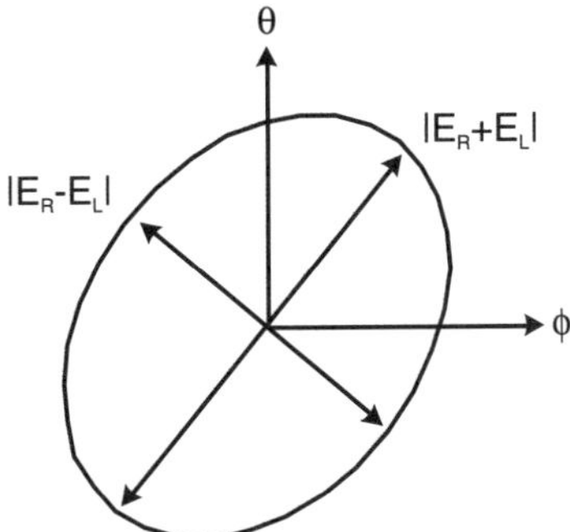

Figure 2.17 Definition of elliptical polarization.

The cross-polarization ratio of a circularly polarized field is defined with respect to E_R as:

$$\mathrm{XPD} = 20 \log \left| \frac{E_R}{E_L} \right| \tag{2.16}$$

A standard antenna used in circular polarization measurements is the loop or helix antenna. The XPR for the standard antenna should be small enough to measure the inversely rotated polarization antenna. For the elliptical polarization case, the axial ratio is defined as the ratio between the maximum and minimum amplitudes as:

$$\mathrm{AR} = 20 \log \left| \frac{E_R + E_L}{E_R - E_L} \right| \tag{2.17}$$

Another common measurement performed for circular polarized antennas is the so-called "spin linear technique" in which a linearly polarized standard antenna is rotated much faster than the azimuth rotation of the DUT. A radiation pattern measured using this method is shown in Figure 2.18. The peak-to-peak level of the ripple in the pattern gives the axial ratio of the DUT.

2.2.3.2 Gain Measurements (Relative Gain, Pattern Integration, and Circular Polarization)

Antenna gain is classified into directive gain and absolute gain. The definition of directive gain does not include input impedance mismatch losses and losses inside the antenna (e.g., ohmic losses). It defines only the directivity of the measured antenna, that is, the radiated energy distribution in space from the antenna. The absolute gain includes all sources of loss for the antenna and is the term most often used when specifying the transmission power and link margin in real operating environments. The absolute antenna gain, G_s, is measured by comparing the received electric field strength between the standard gain antenna and the DUT. An example of the absolute gain of a standard antenna is $G_s = 2.15$ dBi for a half-wavelength dipole antenna. An antenna with this value of gain can be used to measure a DUT with an expected gain, G_d, of less than 15 dBi. A standard high-gain pyramidal horn antenna, however, can be used to measure antennas with an expected gain in the range of 15 to 35 dBi. This criterion is not theoretical but was determined empirically by evaluation of measurement data from relative gain measurements on various types of antenna. Figure 2.19(a) shows a typical

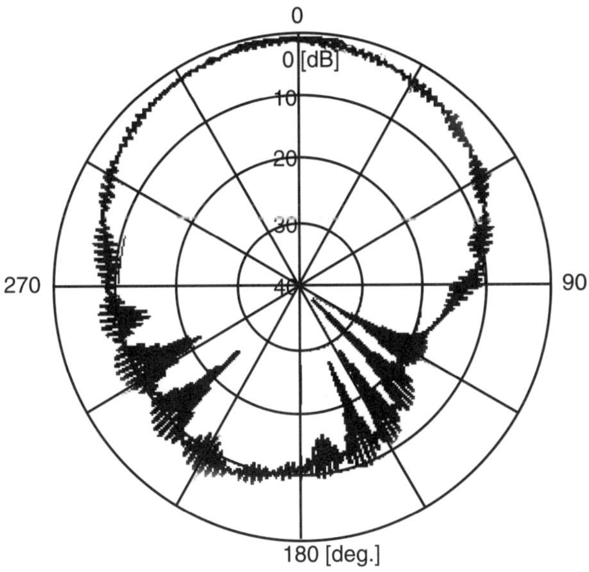

Figure 2.18 Measured spin linear radiation pattern.

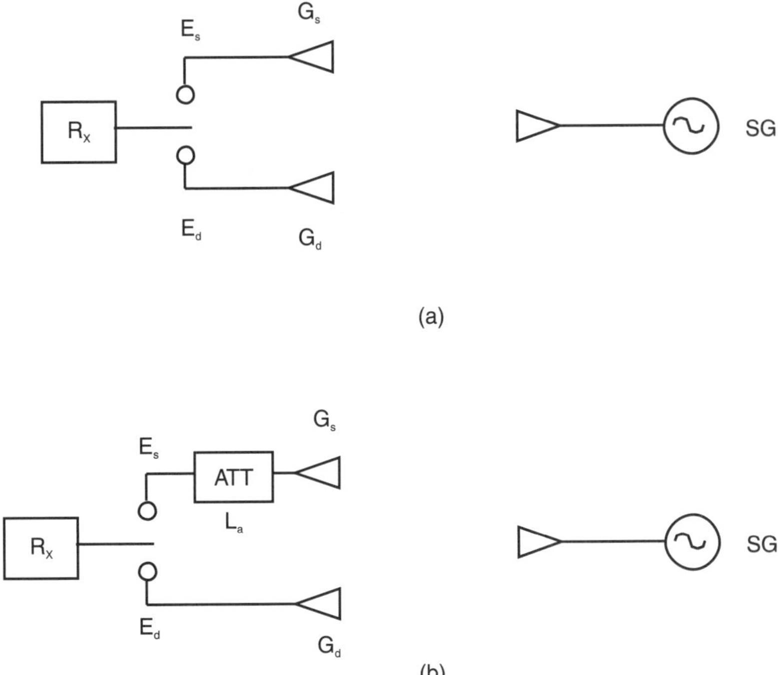

Figure 2.19 (a) Measurement setup for antenna gain and (b) measurement setup for antenna gain by attenuator.

measurement setup. For low-gain antennas with gains of less than 10 dBi, the gain measurement can be performed inside an anechoic chamber.

The received electric field strength E is measured in units of dBμV using a standard antenna with an antenna gain G_s measured in advance. Using the same experimental setup, the test antenna replaces the standard antenna and once again the electric field strength is measured.

The gain of the measured antenna G_d is thus given as:

$$G_d(\text{dB}) = E_d(\text{dB}) - E_s(\text{dB}) + G_s(\text{dB}) \tag{2.18}$$

A more precise gain measurement can be carried out using a calibrated attenuator; the measurement setup is shown in Figure 2.19(b). The attenuator is inserted between the standard antenna and the receiver and is adjusted until the received signal levels from both the standard antenna and the DUT become equal. If the attenuation level is then L_a, the antenna gain can be obtained from:

$$G_d(\text{dB}) = G_s(\text{dB}) - L_a(\text{dB}) \tag{2.19}$$

Measuring the directivity gain by this procedure is referred to as a pattern integration method. The directivity gain in the direction of maximum radiation (θ_o, ϕ_o) is defined by the radiation power density function as:

$$g_p(\theta_o, \phi_o) = 4\pi \frac{p_p(\theta_o, \phi_o)}{\int_0^{2\pi} \int_0^{\pi} \{p_p(\theta, \phi) + p_x(\theta, \phi)\} \sin\theta \, d\theta \, d\phi} \tag{2.20}$$

where $\mathbf{p}(\theta, \phi)$ consists of a principal polarization component $p_p(\theta, \phi)\mathbf{e}_p$ and a cross-polarization component $p_x(\theta, \phi)\mathbf{e}_x$ and where the unit vector for each component is denoted by $\mathbf{e}_p$ and $\mathbf{e}_x$. Figure 2.20 shows the polar coordinate system for the measurement. Measurement of the θ angle is performed in the elevation plane ($0 \leq \theta \leq \pi$) and of the ϕ angle in the azimuth plane ($0 \leq \theta \leq 2\pi$).

When measurement of the antenna radiation pattern for both the principal and cross-polarization components is performed over the whole of space, the directivity gain, as defined by (2.20), is obtained because the power density function is expressed in terms of the directivity as:

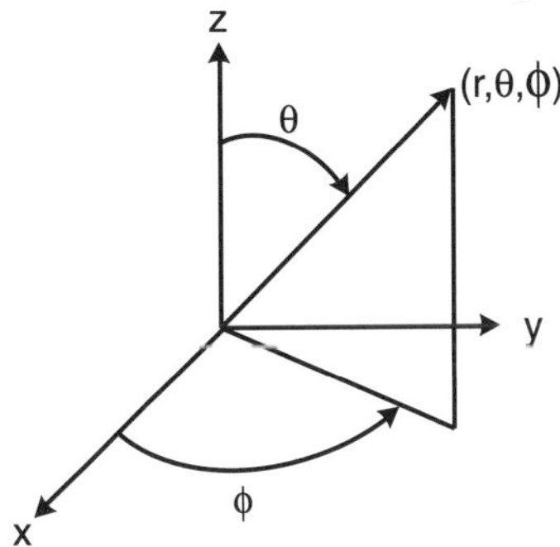

Figure 2.20 Polar coordinate system for the measurement.

$$p_i(\theta, \phi) = C\left|D_i(\theta, \phi)\right|^2, \qquad i = p, x \tag{2.21}$$

where C is the constant of proportionality.

The pattern integration method requires the receiving probe to be moved in a spherical or hemispherical pattern. For the measurement system shown in Figure 2.21(a), the DUT is stationary while the receiving antenna is moved through a semicircle on the holding frame. The plane of the measurement is gradually shifted through 180° until a full hemisphere is covered. On the other hand, in Figure 2.21(b) the receiving antenna is moved through a quarter circle while the DUT is rotated through 360° in order to cover a full hemisphere. In Figure 2.21(b), the semicircular frame on which the receiving antenna is mounted is placed on a ground plane in order to use the image of the DUT radiation pattern. Otherwise, a microwave absorber must cover the floor on which the DUT is mounted to suppress multipath reflections [Figure 2.21(b)]. Note, however, that if the image method of measurement is to be adopted (described in Section 2.1.3), the DUT must be physically symmetrical about a given plane.

For the measurement system shown in Figure 2.21(b), the test antenna must be scanned twice in each plane to measure the normal and cross-polarizations of the DUT pattern by rotating the receiving antenna by 90°. This process then gives data for a semicircle in one plane of measurement. Then the same measurement is performed after resetting the DUT to cover the spherical surface. This method of scanning creates a spatial map of the radiation field of the DUT in the azimuth and elevation directions, M and

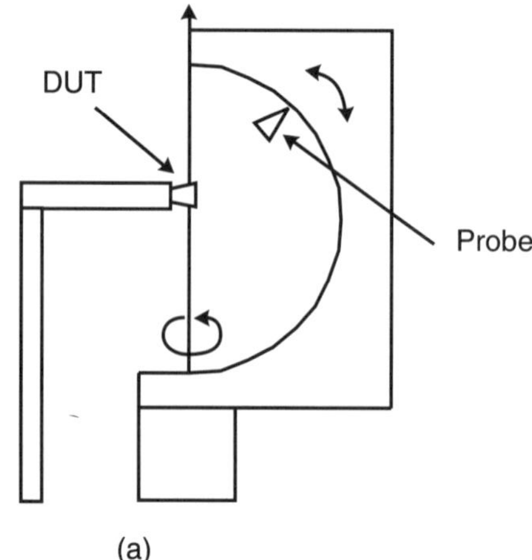

(a)

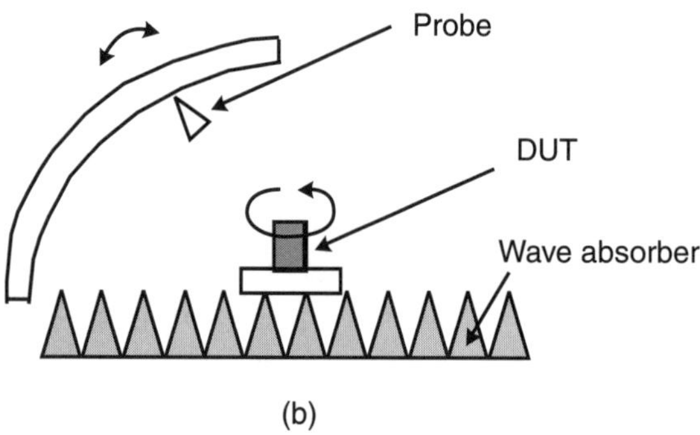

(b)

Figure 2.21 (a) Spherical scanning and (b) hemispherical probe scanning.

N, respectively. The directivity gain may then be obtained directly from (2.20) and (2.21) as:

$$g_p(\theta_o, \phi_o) = 4\pi \frac{\left|D_p(\theta_o, \phi_o)\right|^2}{\sum_{i=1}^{M}\sum_{j=1}^{N}\left\{\left|D_p(\theta_i, \phi_j)\right|^2 + \left|D_x(\theta_i, \phi_j)\right|^2\right\} \sin\theta_i \Delta\theta\Delta\phi} \tag{2.22}$$

where and $\Delta\theta = 2\pi/M$ and $\Delta\phi = 2\pi/N$.

2.3 Radiation Efficiency Measurements

The demand for miniaturized portable handset designs in mobile communication systems has placed severe restrictions on the size of antennas. It is well known that electrically small antennas suffer from a deterioration in the antenna characteristics including reduced frequency bandwidth, radiation efficiency, and antenna gain. The term *electrically small* refers to an antenna whose length is much smaller than the operating wavelength. The radiation efficiency of such antennas rapidly decreases with a reduction in the antenna size. For portable pagers operating in the VHF band, this is often a serious problem. The radiation efficiency of an antenna is thus an important parameter to be considered in the design and measurement of new antennas for mobile communications. In this section the radiation efficiency is defined and several measurement techniques appropriate to small antennas are described.

2.3.1 Definition of Radiation Efficiency

Losses in antennas, R_l, include ohmic losses due mainly to losses in the conductor, and dielectric losses in the dielectric material of the antenna, if present. If the radiation power from the antenna is termed the radiation resistance, R_r, then the radiation efficiency is defined as:

$$\eta = \frac{R_r}{R_r + R_l} \tag{2.23}$$

2.3.2 Wheeler Cap Method

The antenna input impedance value includes terms for radiation resistance and ohmic losses. If only the ohmic losses are accounted for in the measurement, the radiation efficiency can be calculated according to (2.23). H. A. Wheeler proposed a method to measure the ohmic losses of small antennas by covering the antenna with a small conducting cap, as shown in Figure 2.22 [7, 8]. In this measurement, the antenna should be mounted above a ground plane.

The original method suggested the use of a cap with a radius, a, where a must satisfy the condition $k_o a < 1$, and $k_o = 2\pi/\lambda$ is the wave number in free space. This criterion is based on the radiation characteristics of an infinitely small current element such that the reactive power not related to the radiation is the dominant component in the vicinity of the antenna. However, experimental data from several sources indicate that it is not necessary

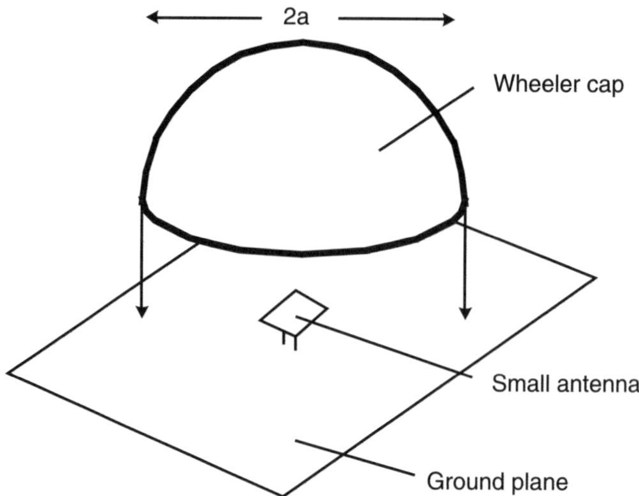

Figure 2.22 Wheeler cap method.

for the spherical cap to adhere closely to this criterion [9]. It is only necessary for the cap length to be less than a half wavelength so as not to excite resonance effects at the cap. An alternative cap shape to the hemispherical cap is the box type [10].

Current flow on the antenna is not affected by the cap, which only affects the ohmic and dielectric losses in the antenna. The measured resistance of the antenna covered by the cap is equivalent to the losses in the antenna, while the real part of the measured input impedance of the antenna without the cap includes the radiation and losses. Thus, using these two measured values, it is possible to find the radiation efficiency of the antenna using (2.23). A theoretical evaluation of the Wheeler cap method shows that this method is of high accuracy for antennas with radiation efficiencies of less than 90%.

Any resonances between the antenna and the cap can cause increased measurement error. The measurement accuracy can thus be a function of the frequency and any resonances present in the measurement will appear as a small dip or spike in the measurement curve. The frequency characteristic of the measurement can therefore give an indication of possible resonances in the experimental setup.

2.3.3 Pattern Integration Method

If the current on the antenna is denoted by I, the radiation efficiency in (2.23) can be rewritten as:

$$\eta = \frac{R_r I^2}{R_r I^2 + R_l I^2} = \frac{P_r}{P_r + P_l} \tag{2.24}$$

where P_r and P_l are the radiation power and power loss of the antenna, respectively. The denominator in (2.24) is equal to the antenna input power and can be measured by a power meter connected to the transmitter power output port. The radiation power P_r can be obtained by measuring the far-field pattern of the test antenna using a receiving antenna of known gain G_r. This measurement procedure is the same as that described in Section 2.2.2.

The radiation power is defined as:

$$P_r = \frac{4\pi R^2}{\lambda^2 G_r} \int_0^{2\pi} \int_0^{\pi} \{p_p(\theta, \phi) + p_x(\theta, \phi)\} \sin\theta \, d\theta \, d\phi \tag{2.25}$$

and this equation is identical to the definition of antenna gain shown in (2.22). The distance R represents a sphere of radius R centered on the antenna under test (DUT). This measurement may be made very precisely because the measurement procedure follows the mathematical definition of the radiation efficiency.

2.3.4 Relative Gain Measurement (Random Field Measurement) Method

The random field measurement (RFM) method is an experimental method for measuring the antenna gain that does not require the use of a large-scale anechoic chamber [11]. The measurement procedure is the same as that for the attenuation coefficient measurement described in Section 1.2.2. A short measurement range is adopted for this method and a measurement site such as an area around a tall building may be selected, as was described in Section 1.2.5. The receiver is located far from the building at a distance of a few hundred meters or more and makes use of multiple scattered signal paths from the transmitter to the receiver (mobile terminal). The mobile terminal is moved along a predefined route first using a standard antenna of gain G_s; a standard antenna such as a dipole antenna is suitable for this type of measurement. The received signal strength (P_s) is recorded and used as a reference signal for the next measurement involving replacement of the standard antenna by the DUT. The DUT on the mobile terminal is then moved along the same predefined route as before and the received signal strength (P_{DUT})

is once again recorded. The relative gain can thus be calculated by taking the ratio of P_{DUT} to P_s. This measurement procedure is the same as that adopted for short-range propagation measurements where the median of the signal is adopted as the standard antenna received power level.

$$G_{DUT} = \frac{P_{DUT}}{P_s} + G_s \tag{2.26}$$

If the measurement is performed with an operator holding the handheld terminal, the effects of the human body on the measurement must also be taken into account. This measurement is thus effective for testing handheld terminal antennas such as those used on mobile phones.

2.4 Diversity Characteristics Measurements Using Radiation Patterns

To reduce the fading effect in the real propagation environment, diversity reception as described in Section 1.3 is used for mobile terminals. The diversity performance is obtained by means of the propagation measurement in Section 1.5.1, but it is also evaluated by measuring the phase and amplitude pattern (complex radiation pattern) of the antenna. This section presents the definition of diversity measurement using a complex pattern and its measurement procedure.

2.4.1 Definition of Correlation Coefficient Using a Complex Pattern

Diversity performance can be characterized by the correlation coefficient (1.16) from Chapter 1. If a two-antenna space diversity system is considered with an antenna spacing of d (Figure 2.23), the received electric fields from antennas e_1 and e_2, respectively, can be expressed as follows:

$$e_1 = \sum_{i=1}^{N} E_i G_{1i} \exp(j\phi_i) \tag{2.27}$$

$$e_2 = \sum_{i=1}^{N} E_i G_{2i} \exp(j\phi_i - k_o \mathbf{r} \cdot \mathbf{d}) \tag{2.28}$$

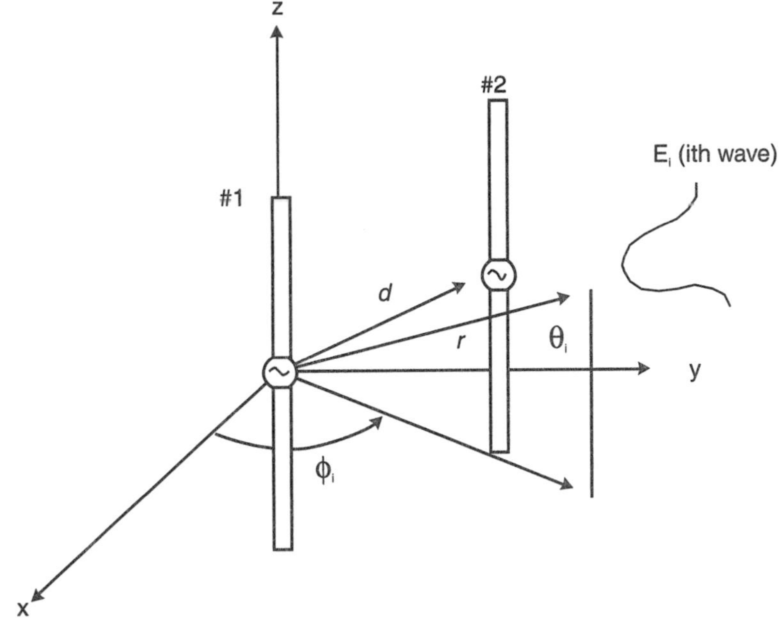

Figure 2.23 Coordinate system for space diversity.

These equations assume that N waves are incident on the antenna with an amplitude E_i and a phase ϕ_i, where G_{1i} and G_{2i} are the antenna gain figures in the direction of the ith incident wave, and k_o is the wave number in free space. The vectors **r** and **d** are, respectively, the unit direction vector for the incident wave E_i and the location vector of antenna number 2 relative to antenna number 1, which is located at the origin of the coordinate system.

For the case where the phase of the incoming wave is changing randomly with time and the incoming waves all have the same amplitude, the mean values of e_1 and e_2 become zero, as $\langle e_1 \rangle = \langle e_2 \rangle = 0$. Hence, the numerator in (1.16) is given by:

$$\frac{1}{2}\left\langle (e_1 - \langle e_1 \rangle)^* (e_2 - \langle e_2 \rangle) \right\rangle$$

$$= \frac{1}{2} \sum_{i=1}^{N} E_i^* G_{1i}^* \exp(-j\phi_i) \sum_{i=1}^{N} E_i G_{2i} \exp(-j\phi_i - k_o \mathbf{r} \cdot \mathbf{d}) \qquad (2.29)$$

The ith incoming wave from the (θ_i, ϕ_i) direction can be expressed by the delta function $\delta(x)$ as:

$$E_i = \frac{\delta(\theta - \theta_i)\delta(\phi - \phi_i)}{\sin\theta} \tag{2.30}$$

where $\sin\theta$ is a metric coefficient in the spherical coordinates. Then, with $E_i^* E_j = 0$, $i \neq j$ given by (2.30), (2.29) can be expanded as:

$$\frac{1}{2}\left\langle (e_1 - \langle e_1 \rangle)^* (e_2 - \langle e_2 \rangle) \right\rangle = \frac{1}{2}\sum_{i=1}^{N} E_i^* G_{1i}^* E_i G_{2i} \exp(-k_o \mathbf{r} \cdot \mathbf{d}) \tag{2.31a}$$

Using the same procedure, the denominator in (1.16) can be expressed as:

$$\frac{1}{2}(e_1 - \langle e_1 \rangle)^2 = \frac{1}{2}\sum_{i=1}^{N} E_i^2 G_{1i}^2, \quad \frac{1}{2}(e_2 - \langle e_2 \rangle)^2 = \frac{1}{2}\sum_{i=1}^{N} E_i^2 G_{2i}^2 \tag{2.31b}$$

If the same amplitude is assumed for all incoming waves, the correlation coefficient ρ is given in [12] as:

$$\rho = \frac{\frac{1}{2}\left\langle (e_1 - \langle e_1 \rangle)^* (e_2 - \langle e_2 \rangle) \right\rangle}{\frac{1}{2}\sqrt{\left\langle (e_1 - \langle e_1 \rangle)^2 (e_2 - \langle e_2 \rangle)^2 \right\rangle}} = \frac{\langle e_1^* e_2 \rangle}{\sqrt{e_1^2}\sqrt{e_2^2}}$$

$$= \frac{\sum_{i=1}^{N} E_i^* G_{1i}^* E_i G_{2i} \exp(-k_o \mathbf{r} \cdot \mathbf{d})}{\sqrt{\sum_{i=1}^{N} G_{1i}^2}\sqrt{\sum_{i=1}^{N} G_{1i}^2}} \tag{2.32}$$

This equation can be converted to an integral form using the distribution function of the incoming waves in the θ and ϕ directions, which are represented by $P_\theta(\theta, \phi)$ and $P_\phi(\theta, \phi)$, respectively. The cross-polarization ratio $\text{XPR} = P_h/P_v$, as defined in (1.15), can also be included so that (2.32) becomes:

$$\langle e_1^* e_2 \rangle = P_h \int_0^{2\pi} \int_0^{\pi} \{\text{XPR}\, G_{1\theta}^* G_{2\theta} P_\theta + G_{1\phi}^* G_{2\phi} P_\phi\}$$
$$\exp(-k_o \mathbf{r} \cdot \mathbf{d}) \sin\theta \, d\theta \, d\phi \qquad (2.33a)$$

$$\sqrt{e_k^2} = P_h \int_0^{2\pi} \int_0^{\pi} \{\text{XPR}\, G_{k\theta}^* G_{k\theta} P_\theta + G_{k\phi}^* G_{k\phi} P_\phi\}$$
$$\exp(-k_o \mathbf{r} \cdot \mathbf{d}) \sin\theta \, d\theta \, d\phi, \quad k = 1, 2 \qquad (2.33b)$$

where G_{km} is the antenna gain pattern for the kth antenna in the $m = \theta$ or ϕ directions.

Using (2.33a) and (2.33b), the correlation coefficient for a two-branch diversity antenna can be evaluated by measuring the amplitude and phase of the radiation pattern for each antenna individually over the whole of space.

2.4.2 Diversity Measurement Procedure

As shown in the previous section, to evaluate the correlation coefficient of a diversity antenna, the complex radiation pattern is first required. The complex radiation pattern can be measured or it can be calculated numerically using computer simulation techniques. For evaluation of the correlation coefficient, however, it is necessary to estimate the distribution function of the incoming waves. Measurement of the incoming wave distribution is not easy to perform since the measurement requires the use of a very high gain antenna such as a parabolic reflector antenna [13].

In mobile communications applications, therefore, an assumption is normally made about the incoming wave distribution. As shown in Section 2.4.1, for the case when the transmitter is far from the receiving position, it is possible to assume that the distribution of incoming waves is uniform in the horizontal plane and has a Gaussian distribution in the vertical plane. The distribution function is then given as:

$$P_m(\theta, \phi) = \frac{C_k}{\sqrt{2\pi\sigma_m^2}} \exp\left\{\frac{(\theta - \theta_m)^2}{2\sigma_m^2}\right\}, \quad m = \theta, \phi \qquad (2.34)$$

where θ_m is the mean elevation angle of the mth incoming wave from the horizontal direction, and σ_m is the standard deviation of mth wave distribution in the elevation plane. Typical examples of these values are $\theta_m = 0$ and $\sigma_m = 40°$ for the case of measured data in an urban area [14].

By neglecting the distribution function in the elevation plane, the simplest approximation for (2.34) can be given by the following delta function:

$$P_m(\theta, \phi) = \frac{\delta(\theta)}{\sin\theta} \tag{2.35}$$

If this particular assumption can be adopted, then the radiation pattern from the antenna need only be measured in one plane and the data can be used to evaluate the correlation coefficient.

2.5 EMC Measurements

In the evaluation of handheld communications terminals, it is important to measure both the main radiation power and the spurious radiation power from the terminal. The conventional measurement method is to evaluate separately the radiation characteristics of the antenna and the terminal standing alone with no antenna connected. However, for a realistic evaluation of the communications terminal, it is necessary to know the combined frequency characteristics of the terminal when joined to the antenna.

The first part of this section describes EMC measurement methods for measuring the radiation characteristics of the terminal and antenna together. These include using an open site direct wave method for the evaluation as well as the previously mentioned random field measurement method used for evaluating the terminal performance in a real environment. For these types of measurements, both a large measurement area and a long measurement time are needed. To reduce both the space and time required when performing the measurements, several measurement systems that are suitable for use in confined spaces are proposed. The second part of the section is devoted to a description of EMC evaluation procedures using compact anechoic boxes/chambers.

2.5.1 Electromagnetic Interference Measurement

All electronic equipment radiates unwanted electromagnetic waves that can be a potential cause of interference to other devices. Each country has a standard for the acceptable level of electromagnetic interference (EMI) that a device may emit. These national standards are based on the International Special Committee on Radio Interference (CISPR) published regulatory standards [15]. The frequency range of the EMI measurements that concern us in this book is from 30 MHz to 1 GHz. A standard open site measuring facility used for conducting such measurements is shown in Figure 2.24. In this system the direct wave from the device under test (DUT) as well as the wave reflected via the conducting ground plane are measured together by the receiving antenna. In indoor testing facilities, a conducting ground plane is also used to ensure a good reflection from the ground. Depending on the type of DUT under test, the distance D from the DUT to the receiving antenna may be 3, 10, or 30m. For the case of the open site test facility, it is important to ensure that there are no other spurious reflections from scattering objects inside the measurement area, which is of length $2D$ and width $\sqrt{3D}$. The performance of the site is evaluated by the site attenuation factor. The definition of this factor is similar to that for propagation loss described in Section 1.1.

Assume that the received signal is the receiving voltage V_d when the transmitting cable is directly connected to the receiving cable with the DUT and receiving antenna removed. The height of the transmitting antenna is fixed at H and the receiving antenna height, h, is changed until the maximum receiving voltage V_r is found. The site attenuation factor is then defined as:

$$\mathrm{SA} = V_d + V_r - (\alpha_t + \alpha_r) \tag{2.36}$$

where α_t and α_r are the feeder losses at the transmitting and receiving antennas. All measurements are in units of decibels. An acceptable error for

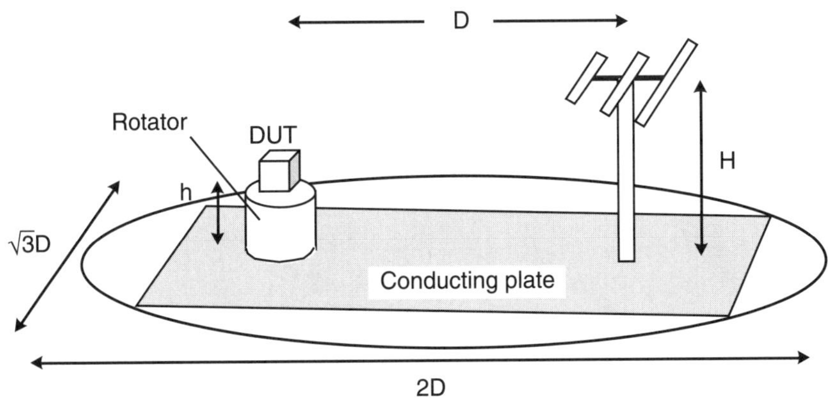

Figure 2.24 EMI measurement using open site.

the site attenuation is ±4 dB compared with the expected theoretical value. This error value indicates the difference between an ideal site and a real test site. The measurement error for the test site described in [16] is less than 0.3 dB.

To simplify the measurement procedure, the DUT is mounted on a rotating table, thus enabling the field radiated by the DUT to be measured over a full 360° circle. To ensure good constructive wave interference between the direct and reflected waves from the DUT, the receiving antenna height (h) is adjusted until the maximum receiving voltage V_m is found. For distances within the range of $D = 3$ to 6m, h is in the range of 1 to 4m, whereas for a distance $D = 30$m, h is found to be between 2 and 6m. The angular position of the DUT relative to the receiving antenna is also adjusted until a maximum V_m is found. For this situation the incoming electric field level E is calculated using of the antenna factor (AF) as:

$$E = \mathrm{AF} \times V_m \tag{2.37}$$

The antenna factor of a half-wavelength dipole antenna is equivalent to the effective length λ/π of the antenna. However, it must be remembered that it is important to include cable and feeder connector losses in the measurement. In actual measurement systems the antenna factor is evaluated by comparing the antenna under measurement with a calibrated standard antenna. The AF is then given in terms of the calibrated antenna factor AF_o and the received voltage V_o for the standard antenna as

$$\mathrm{AF} = \frac{\mathrm{AF}_o \times V_o}{V_m} \tag{2.38}$$

In this case, the antenna factor is only defined as a constant value for the horizontal polarization component, since the radiation pattern of the vertical dipole changes with the elevation angle and the AF becomes a function of height. To exclude the antenna factor from the site attenuation, the normalized site attenuation is defined below:

$$\mathrm{NSA} = V_d + V_r - (\mathrm{AF}_t + \mathrm{AF}_r) - \Delta\mathrm{NSA} \tag{2.39}$$

where AF_t and AF_r are the antenna factors of the transmitting and receiving antennas, respectively, and ΔNSA is a correction factor for mutual coupling between the two antennas and the effect of the Earth's surface [17, 18].

Depending on the frequency range of the EMI measurement, the type of antenna used differs [19]. For example, in the 30- to 1000-MHz range, a half-wavelength dipole antenna or a log-periodic dipole array can be used; for the 30- to 300-MHz range, a biconical antenna; and for 1 GHz and above, a corrugated horn antenna.

2.5.2 Reverberating Enclosure Measurement

The open site measurement method described in the preceding section is time-consuming because the DUT must be rotated in order to scan the full 2-D or 3-D radiation patterns. Additionally, the height of the receiving antenna must be adjusted until the maximum signal level is found. An alternative measurement system for EMI measurements is proposed that uses a reverberating chamber [20] (Figure 2.25). The measurement system consists of two flat vanes installed on adjacent walls of the chamber with the transmitting (TX) and receiving (RX) antennas mounted on the remaining two walls of the chamber. The vanes are rotated at speeds of 90 and 120 rpm (revolutions per minute), respectively. The differing speeds of rotation generate a random field inside the chamber. The chamber size for this particular experimental setup is 2m × 2m × 2m with the DUT mounted on a fixed pedestal in the center of the chamber. The frequency range for this size of chamber is above 1 GHz.

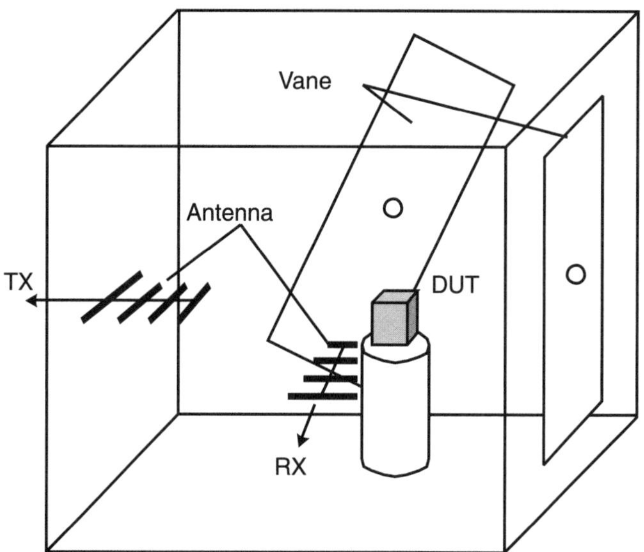

Figure 2.25 EMI measurement using reverberating chamber.

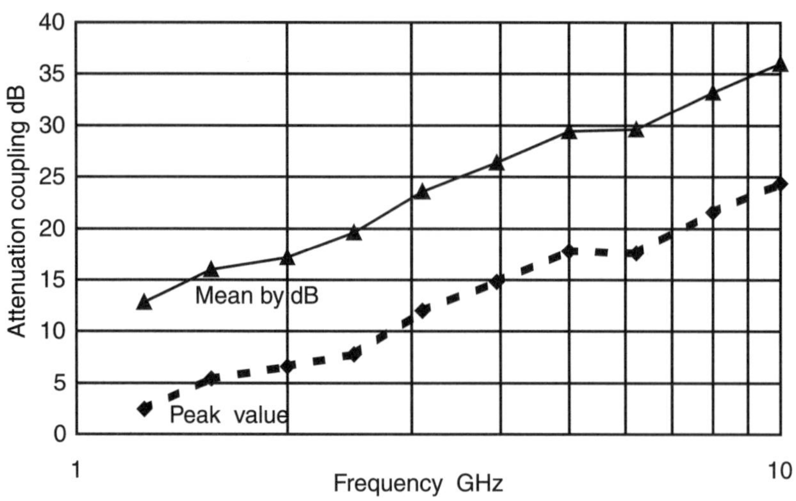

Figure 2.26 Coupling attenuation inside reverberating chamber.

The transmitting and receiving antennas are broadband log-periodic antennas for the frequency range of 1 to 10 GHz. An example of the site attenuation as a function of frequency is shown in Figure 2.26, the mean value being calculated in units of decibels. The rate of collection of data points is 10 per minute. The measurement method generally has an acceptable error of around a few decibels. The advantage of the reverberating system is the short time required to perform the measurement. The disadvantage, however, is an increase in the measurement error to a few decibels.

2.5.3 Random Field Measurement

Random field measurement has already been presented in Section 2.3.4 as an outdoor measurement method for antenna gain. It is now presented in this section as an indoor EMI measurement method. The radiation pattern of a mobile handheld terminal can be seriously distorted from the ideal by the size of the casing and also by the presence of the operator's hand, head, and body. To counteract these effects, the radiated power from the terminal should be randomly scattered by obstacles, as would be experienced when operated in a real environment. In an outdoor environment, this test method requires only a suitable transmitting and receiver system and an appropriate site for testing. However, the uncertainty of weather conditions from day to day and also the need to obtain a testing license for the use of a high-power transmitter outdoors, makes outdoor testing of the mobile terminal troublesome. To overcome these obstacles, a random field measurement system has been developed for indoor usage. The indoor testing method is also good as a first test of the DUT before moving on to the final outdoor test.

The measurement system is placed inside a shielded room and many series of scatterers are hung from the ceiling as shown in Figure 2.27 [21]. In order to generate random field inside shield room, the scatters surround the antenna under test, and the receiving antenna is mounted on a rotating pedestal throughout the measurement. The electromagnetic wave scatter obstacles are made from aluminum-covered cardboard strips bent into a parabolic-like cross-section. The strip length is a quarter-wavelength at measurement frequency. The receiving antenna is rotated simultaneously in two directions: in the horizontal plane by the inverted L-shaped rotor arm mounted on the pedestal and in the vertical plane by a smaller rotor arm mounted perpendicularly to the first rotor arm. The measurement procedure is the same as that described in Section 2.3.4. In the first step, a standard half-wavelength dipole antenna of known gain characteristic is connected to the transmitter and mounted on the fixed pedestal (shown on the right side of Figure 2.27). Using this setup, the median value of the received level M_s is measured by rotating the receiving antenna (left side of Figure 2.27). Next, the same measurement is performed on the mobile terminal and the medium value M_t is obtained. The radiated power P_t from the terminal under test can thus be obtained using the relation:

$$P_t = \frac{M_t}{M_s} P_o \tag{2.40}$$

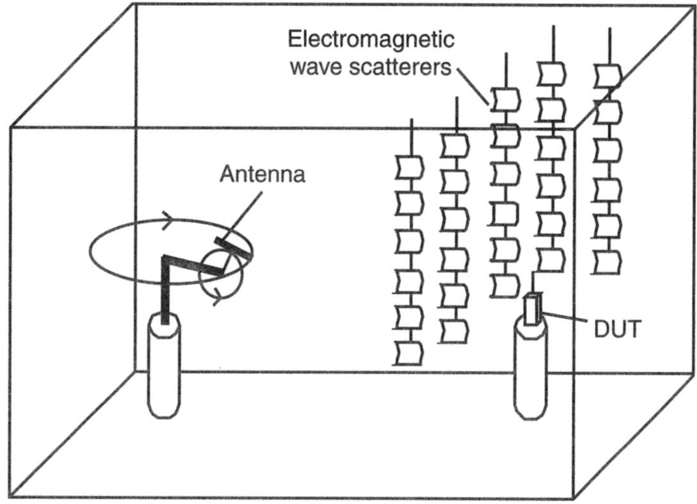

Figure 2.27 Random field measurement using electromagnetic wave scatterers.

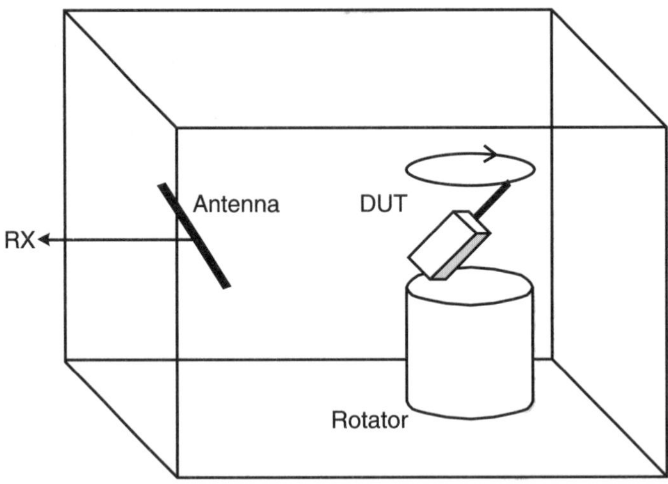

Figure 2.28 Radiation power measurement by compact shield box.

where P_o is the transmitter output power. If the measurement is performed with the operator holding the test terminal, the measured radiation power includes the power loss in the body tissue.

In situations where test space is even more severely restricted, a measurement system using a compact shield box can be used [22]. Figure 2.28 shows the measurement setup. In this measurement system a random electromagnetic field distribution is generated inside the box by means of the reflections off the inner box walls. The handheld terminal is mounted on a rotating pedestal offset from the rotation axis in order to reduce the effects of the terminal antenna directivity on the measurement. The size of the shield box is only 50 cm × 50 cm × 100 cm, roughly 1/30 the size of chambers used in other methods. Its operating frequency is above 1 GHz. The upper limit of the operating frequency depends on the frequency band of the receiving antenna. The receiving antenna is a standard half-wavelength dipole antenna mounted with the element tilted at a 45° angle to the vertical. The DUT is also mounted at a 45° angle on the rotating pedestal.

The tilt angle is used to suppress the direct path signal from the transmitter to the receiver. The measurement error for this method can be less than 1 dB and a measurement time of only a few minutes allows data acquisition of thousands of data points at each frequency.

2.6 Product Testing

Antenna performance measurement procedures are described in this final section of the chapter. In the final development stages of a mobile terminal, several tests are carried out to check the terminal performance under practical operating conditions. This section details examples of testing procedures for the final development and mass production stages of mobile handset terminals.

The testing procedures can be categorized into electrical tests and mechanical tests. Mechanical testing of the handset is not directly relevant to electrical engineers, but it is an important factor in the design of the handset. In this discussion, only digital system mobile terminal handsets are considered. At this stage of handset development, the prototype handset has already been fabricated and tested and the final assembly of the antenna device is the main consideration. Important evaluation tests required for the development of the prototype handset antenna were dealt with in the preceding sections. Similar evaluation test procedures for base station antennas are presented in Chapter 4.

2.6.1 Electrical Testing at the Development Stage

2.6.1.1 Equivalent Radiated Power Measurement (Transmission Test)

The radiated power level and the radiation pattern from the RF transmitter of the mobile handset are measured in an RF anechoic chamber. The radiation pattern for the handset is measured both with the antenna extended from the handset casing and also when the antenna is retracted within the casing. The frequency measurement is made at the center of the operating frequency bandwidth of the device and also at the upper and lower frequency limits. The acceptable VSWR level for the system determines the upper and lower frequency limits.

To ensure that the test data are as realistic as possible, a prototype handset is used for the test measurement. The prototype handset is set to radiate an RF burst signal continuously during the test. The radiation pattern from the device is then measured in the usual manner. The pattern is also measured for the handset with a phantom instead of a human operator. Categorization of phantoms is covered in the next chapter.

2.6.1.2 Received Signal Strength Indicator Measurement (Reception Test)

The prototype handset undergoing testing is placed inside an anechoic chamber together with a base station emulator attached to a transmitter antenna of known gain. For digital cellular communications systems the handset acts as a terminal to report the received signal strength indicator (RSSI) level of the communications uplink in an ideal operating environment. The RSSI pattern for the system can then be measured as a function of the elevation and azimuth angles. The pattern averaging procedure that allows the handset antenna gain to be evaluated is described in the next chapter.

2.6.1.3 Ratio Testing of the Communication Data Link

The prototype handset is tested in a situation in which it is immersed in a weak base station transmitting signal field, in an outdoor environment and the number of successful connections to the base station is logged against the actual number of calls made. This ratio then gives a rough figure of the percentage of successful connections that can be expected for the system under a specific set of operating conditions. This is one of the final tests made in the development of mobile communications terminals. If the ratio is below the target ratio expected for the system, it may be possible to trace the problem to a number of areas in the mobile terminal design, for example, faulty antenna design, bad control software design, or too large an internal noise source from a specific component used in the internal circuit design.

2.6.2 Electrical Tests in the Mass Production Process

Testing of the mobile handset during the mass production process is as important as the many tests performed during the development stages of the product. It is also a key factor in monitoring the quality control level of the final product. At this stage, the transmission power level can only be checked in a very simple fashion. The criterion as to whether or not the final product is viable is whether the measured transmitting power from the handset is above a level specified by the manufacturer. This criterion is set to satisfy the relevant regulations.

2.6.3 Mechanical Testing in the Development Phase

The purpose of mechanical testing of the mobile terminal antenna is to allow the manufacturer to ensure product durability. Again, the criterion for passing the tests is whether or not the transmission power increases above a level specified by the manufacturer. These are the only mechanical tests required for the antenna in the final development phase and they are illustrated in Figure 2.29.

2.6.3.1 Extraction Test

For the extraction test, the whip antenna is pulled in and out of the casing tens of thousands of times. After this test is completed, the electrical characteristics of the antenna are remeasured to verify whether or not the antenna is still within the operating characteristic range of acceptance.

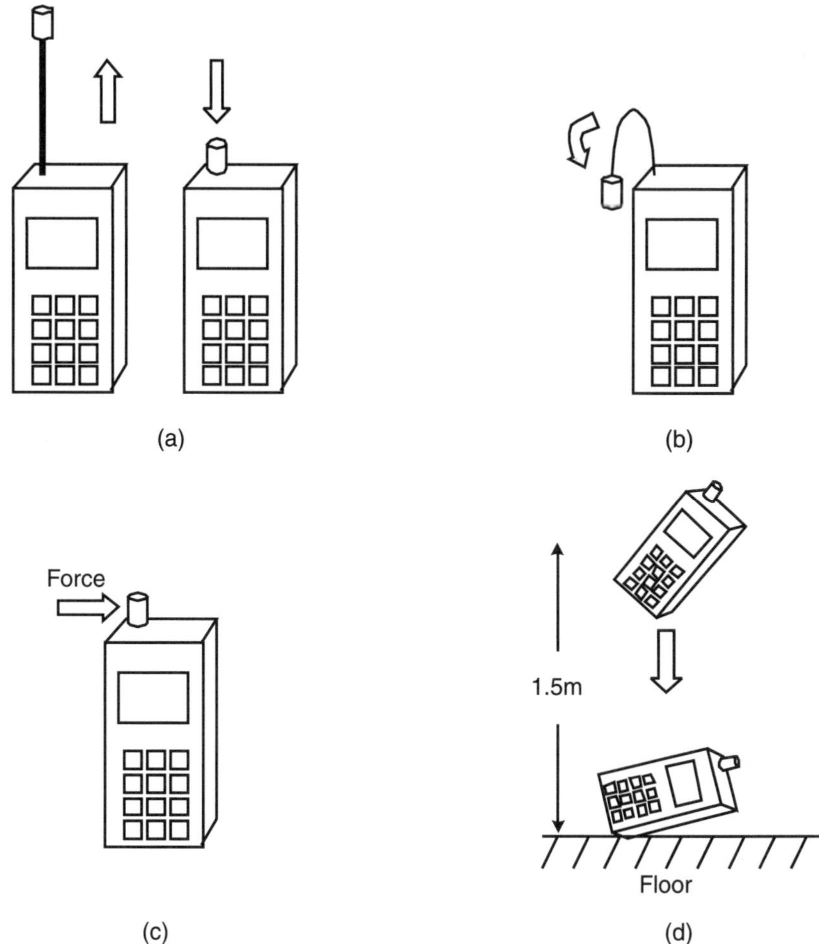

Figure 2.29 Mechanical tests during the development stage: (a) pulling-out test, (b) bending test, (c) load test, and (d) fall test.

2.6.3.2 Bending Test

For this test the whip antenna is bent until the tip touches the chassis of the handset. The whip antenna should have sufficient flexibility that it does not break during this test and should also spring back to its original shape.

2.6.3.3 Load Test for the Retracted Antenna

With the antenna retracted into the handset casing, a force is applied to the tip perpendicular to the antenna. After the test has been performed the antenna performance is rechecked.

2.6.3.4 Fall Test

As a final test, the handset terminal is dropped from a height of around 1.5m and checked to see whether or not it has been broken. This height is roughly the height at which the operator uses the handset. This test completes the mechanical tests required for a whip antenna used on mobile phone terminals.

References

[1] Johnk, C. T. A., *Engineering Electromagnetic Fields and Waves,* 2nd ed., New York: John Wiley and Sons, 1988, pp. 621–625.

[2] Johnson, R. C., *Antenna Engineering Handbook,* 3rd ed., New York: McGraw-Hill, 1993, pp. 43-24–43-25.

[3] Yee, K. S., "Numerical Solution of Initial Boundary Value Problems Involving Maxwell's Equations in Isotropic Media," *IEEE Trans. Antennas and Propagation,* Vol. AP-14, 1966, pp. 302–307.

[4] Han, R. F., and J. G. Fikioris, "Impedance and Radiation Pattern of Antennas Above Flat Discs," *IEEE Trans. Antennas and Propagation*, Vol. AP-21, No. 1, Jan. 1973, pp. 97–100.

[5] "IEEE Standard Test Procedures for Antennas," IEEE Std. 149-1979, pp. 20–22.

[6] Gillespi, E. S., "Measurement of Antenna Radiation Characteristics on Far-Field Ranges," in *Antenna Handbook,* Vol. 4, Chap. 32, New York: Chapman & Hall, 1993.

[7] Newman, E. H., P. Bohely, and C. H. Walter, "Two Methods for the Measurement of Antenna Efficiency," *IEEE Trans. Antennas and Propagation*, Vol. AP-23, No. 4, 1975, pp. 457–461.

[8] Wheeler, H. A., "The Radiation Sphere Around a Small Antenna," *Proc. IRE*, Vol. 47, No. 8, 1975, pp. 1325–1331.

[9] Muramoto, M., N. Ishii, and K. Itoh, "Radiation Efficiency Measurement of a Small Antenna Using the Wheeler Method," *Electronics and Communications in Japan,* Part 1, Vol. 79, No. 6, 1996, pp. 93–100.

[10] Ishii, N., and K. Itoh, "A Consideration of the Thin Planar Antenna With Wire-Grid Model," *IEICE Trans. Communications*, Vol. E76-B, No. 12, 1993, pp. 1518–1525.

[11] Andersen, J. B., and F. Hansen, "Antennas for VHF/UHF Personal Radio: A Theoretical and Experimental Study of Characteristics and Performance," *IEEE Trans. Vehicular Technology*, Vol. VT-26, No. 4, 1977, pp. 349–357.

[12] Adachi, A., et al., "Cross-Correlation Between the Envelopes of 900-MHz Signals Received at a Mobile Radio Base Station Site," *IEE Proc.*, Vol. 133, Part F, No. 6, 1986, pp. 506–512.

[13] Taga, T., "Analysis for Mean Effective Gain of Mobile Antennas in Land Mobile Radio Environments," *IEEE Trans. Vehicular Technology*, Vol. VT-39, No. 2, May 1990, pp. 117–131.

[14] Jakes, W. C., *Microwave Mobile Communications*, New York: John Wiley & Sons, 1974, p. 192.

[15] Morgan, D., *A Handbook for EMC Testing and Measurement,* London: Peter Peregrinus Ltd., IEE, 1994, Chap. 2.

[16] Ohtani, A., et al., "Intercomparison of Antenna Calibration Between CRL and NPL(UK)," *Proc. IEICE General Conf.* B-4-71, Mar. 1998 (in Japanese).

[17] Smith, A. A., Jr., "Calculation of Site Attenuation From Antenna Factors," *IEEE Trans. Electromagnetic Compatability,* Vol. EMC-24, No. 3, 1982, pp. 301–316.

[18] Sugiura, A., et al., "An Improvement in the Standard Site Attenuation Method for Accurate EMI Antenna Calibration," *Trans. IEICE Japan*, Vol. E78, No. 8, 1995, pp. 1229–1237.

[19] MacNamara, T., *Handbook of Antennas for EMC,* Norwood, MA: Artech House, 1995, Chap. 5.

[20] Okamura, M., and A. Sugiyama, "Evaluation of the Performance of a Reverberating Enclosure Used for Making the Measurement of Total Radiated Power From Microwave Apparatus Operating in the Microwave Frequency Range," *Proc. Int. Symp. EMC,* No. 18, AA3, 1984, pp. 594–598.

[21] Maeda, T., and T. Morooka, "Experimental Studies and Improvements on the Accuracy of the Indoor Random Field Measurement Method for Obtaining the Radiation Efficiency of Electrically Small Antennas," *IEICE Japan Trans. B.*, Vol. J71-B, No. 11, 1988, pp. 1259–1265.

[22] Arai, H., and T. Urakawa, "Radiation Power Measurement Using Compact Shield Box," *Proc. Int. Symp. Antennas and Propagation*, Chiba, Japan, 1996, 4B-13.

3

Handset Antennas and Influences Due to the Human Body

In real operating conditions, the characteristics of a handset antenna can be influenced by many components used in the handset. Also, due to the fact that recent handsets are becoming more compact, it has now become necessary to take into account the influence of the handset casing. A wire antenna extending from the handset casing is one of the most common types of antenna used in handset terminal designs. However, such antennas are strongly influenced by the handset terminal casing and also by the user's hand and head when operating the terminal. These influences must be taken into account during the design stages of the handset antenna.

This chapter is therefore devoted to discussion of the techniques used to evaluate such influences. It is possible to use a special type of mannequin, called a *phantom,* in the evaluation of the influence of the human body on the antenna characteristics. Measurement methods are described using a phantom in place of an actual human operator and data are presented that compare measurements using a phantom with those using a human operator.

3.1 Human Body Influences on the Handset Antenna

The purpose of these antenna measurements and the inclusion of the data into the antenna design is to help counteract the effects of the casing and the human body on the antenna operating characteristics. Even if the antenna characteristics prove to be fully satisfactory when the antenna is attached to the terminal casing, when an operator actually uses the handset the antenna characteristics can still be adversely affected. It is therefore vital to the final antenna design that the influences of the human body on the antenna characteristics be fully investigated.

A material that is electrically equivalent to human body tissue is used to design a mannequin called a *phantom* for these evaluation tests. Alternatively, an actual person can be used for the measurements. Initially in this chapter, the effect of the casing on the antenna and the effect of the hand that holds the handset are explained. In the following sections, details of different types of phantom are described.

3.1.1 Relationship Between Antenna Type and Casing Size

Monopole antennas are widely used today for portable phone terminals. If a quarter-wavelength monopole antenna is placed on an infinite ground plane, the total antenna length can be regarded as half a wavelength due to the image current on the ground plane. However, because an infinite ground plane does not exist in reality, the antenna must be installed on a finite ground plane. As has already been explained in Section 2.1.4, the input impedance and radiation pattern of the antenna are influenced by the size of the ground plane. Thus, since the handset casing acts as the ground plane for the monopole antennas on mobile terminals, the antenna characteristics can be strongly affected by the casing size.

As shown in Figure 3.1, the casing can be approximated by a metal board of height L and width W. The monopole antenna can be approximated by a wire model of the antenna of length h as shown in Figure 3.1. The method of moments can be used to calculate the radiation pattern from this model in the z-x plane [1]. Current flowing on the outside of the casing is the cause of undesirable lobes in the radiation pattern from the handset. If the casing height, L, becomes larger than the length of the monopole antenna, h, then a large sidelobe will appear in the z-x plane cut of the radiation pattern, as shown in Figure 3.2(a).

Handset terminals for cellular systems often use monopole antennas approximately half a wavelength in length in order to increase the antenna gain. The actual length of the monopole antenna is $3/8\lambda$ or $5/8\lambda$ in order to take account of impedance matching considerations at the antenna feed point. For the sleeve antenna, unlike the monopole antenna, a ground plane is not required in order to achieve resonance. However, sidelobes appear on the radiation pattern when the length of the casing becomes greater than the antenna length, as shown in Figure 3.2(b). These are also due to the influence of the current flowing on the casing surface.

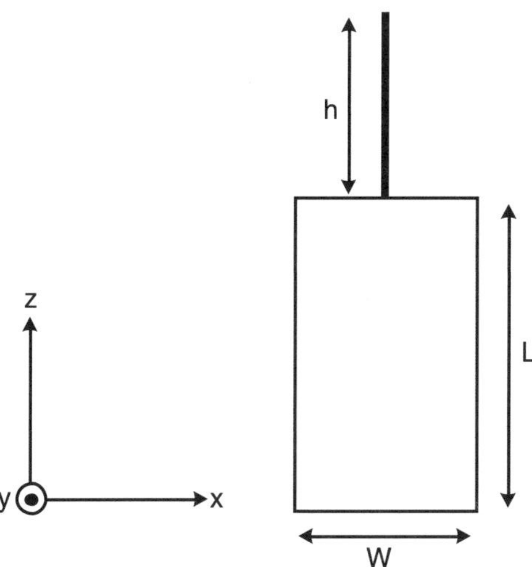

Figure 3.1 Analysis model of an antenna attached to a casing.

By making a notch on the casing, it is possible to partially suppress the current flowing on the casing [2]. To suppress the undesirable radiation more fully, the casing can be completely separated into two parts as shown in Figure 3.3 with the two separate parts of the casing connected by a conducting wire. The resulting radiation pattern for this shape of handset is shown in Figure 3.4. The sidelobes cannot be completely removed using this technique, but handset type (b) does give better sidelobe suppression than handset type (a). If the handset is divided into two isolated sections (with no conducting wire), then the sidelobes in the radiation pattern completely disappear. Thus, a casing length shorter than the antenna is required to completely suppress the current flowing on the casing. When discussing antenna length, it is important to remember that the length refers to the electrical length of the antenna. Use of a helical structure or inductance loading can shorten the physical length of an antenna for mobile terminal applications. Its resonance properties are then equivalent to those of a monopole antenna or a sleeve antenna, and its electrical length is defined as $\lambda/4$ or $\lambda/2$, respectively. Resonance is determined by the current distribution on the antenna or the radiation pattern. When using physically shortened antennas, the casing electrical length must be less than the antenna electrical length.

3.1.2 Relationship Between Average Antenna Gain and Casing Size

As shown in the preceding section, the radiation pattern of the handset antenna is greatly dependent on the relative length of the antenna with respect

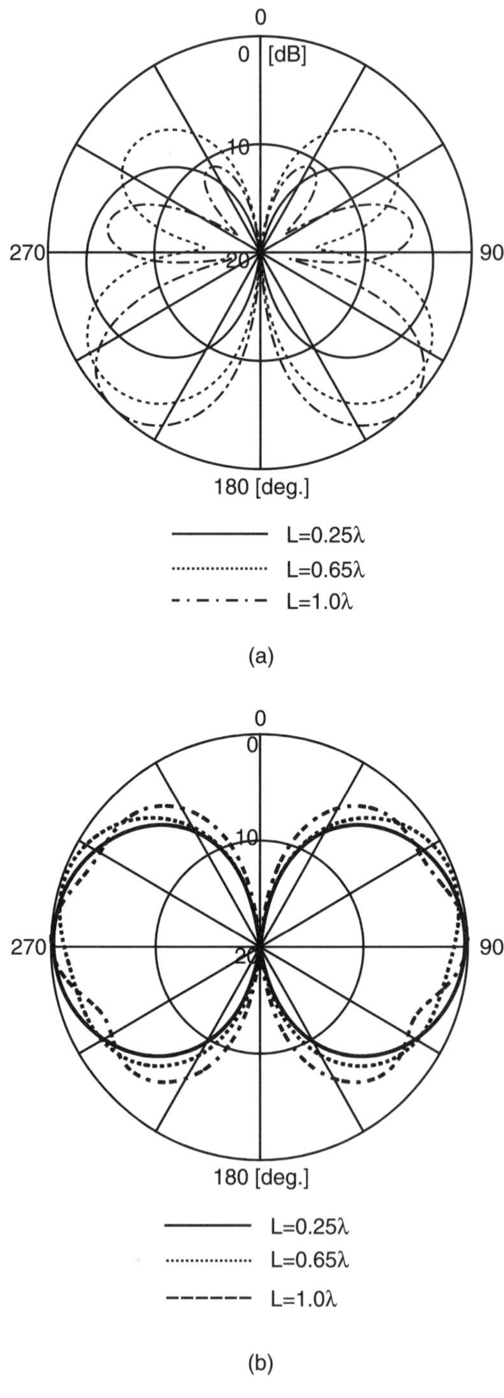

Figure 3.2 Casing size and radiation pattern: (a) $h = 0.25\lambda$ and (b) $h = 0.5\lambda$.

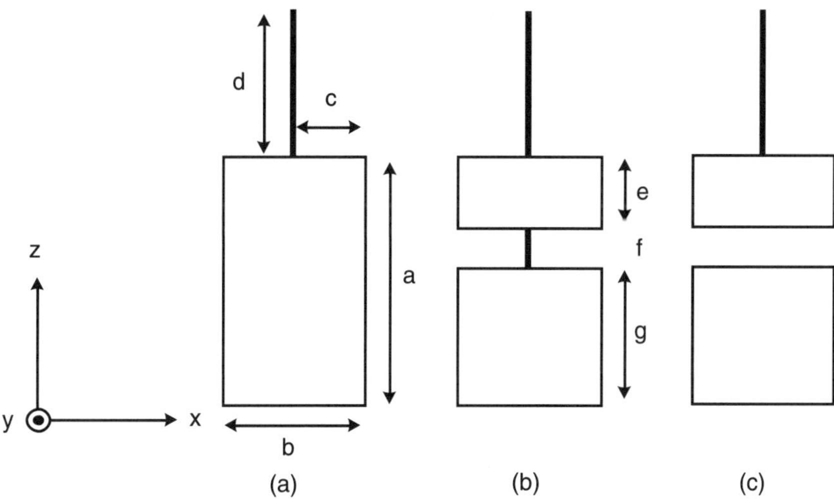

Figure 3.3 Casing shape and radiation pattern, a = 105 mm; b = 48 mm; c = 24 mm; d = 29 mm; e = f = 15 mm; g = 75 mm. (a) Handset model, (b) separated casing connected by a conducting wire, and (c) casing of two separated parts.

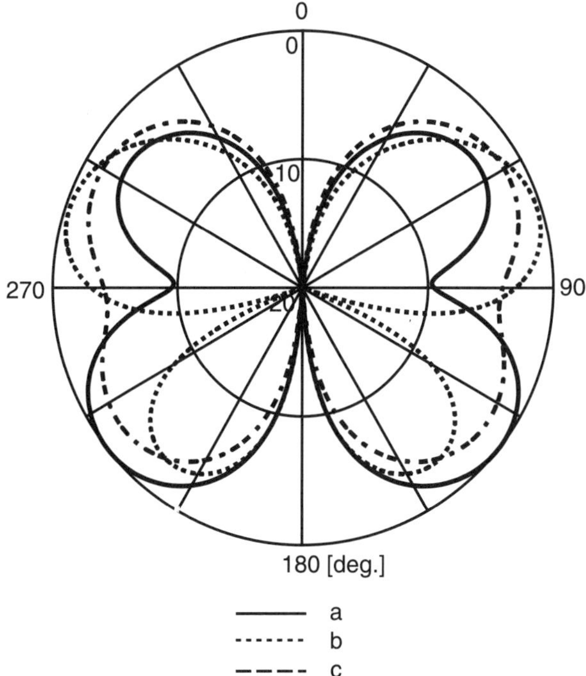

Figure 3.4 Casing size and radiation pattern of sleeve antenna.

to that of the casing. Therefore, it is necessary to make some quantitative estimation of how much radiation pattern distortion will occur if the casing size is changed. The casing width does not change the radiation pattern if the width is less than the casing length, whereas a wide casing increases the frequency bandwidth just like a thick wire antenna. Accordingly, in this section, casing size refers to the length only.

An additional consideration is the slant angle of the handset relative to the operator's head when in actual use. This section shows the influence of the slant angle on the antenna gain and takes into consideration the influence of the head and the hand.

Because electromagnetic waves are scattered by many obstacles when traveling from the base station to the reception point, in modern mobile communication systems designers can assume that the waves will approach the handset from all directions simultaneously. As shown in Section 2.4.1, the waves arriving at the receiving point can be considered to be concentrated in a horizontal plane. The receiving electric field strength E_r is calculated by multiplying the radiation pattern in the horizontal plane of the terminal by the uniformly arriving electric field distribution from the base station. This is then defined as an average received electric field strength. This average received electric field is referred to as the mean effective antenna gain (MEG) when considered from the transmitting antenna viewpoint. E_r is then given as:

$$E_r = \int_0^{2\pi} \frac{f(\theta')}{\sqrt{\sin^2\theta + \cos^2\alpha \sin^2\phi}} \left(\frac{\mathrm{XPR}}{1+\mathrm{XPR}} \sin\alpha + \frac{1}{1+\mathrm{XPR}} \cos\alpha \, \sin\phi \right) d\phi \tag{3.1}$$

where θ' and α are the angles of inclination of the handset as defined in Figure 3.5, and XPR is the cross-polarization ratio (XPR = vertical polarization electric field strength/horizontal polarization electric field strength). The function $f(\theta')$ is given in terms of the E_θ and E_ϕ electric field components, but the E_ϕ can be neglected for most handset terminals. In the handset coordinate system shown in Figure 3.5(a), ϕ and θ' satisfy the following equation:

$$\theta' = \left(1 - \frac{2\alpha}{\pi}\right)|\phi| + \alpha \tag{3.2}$$

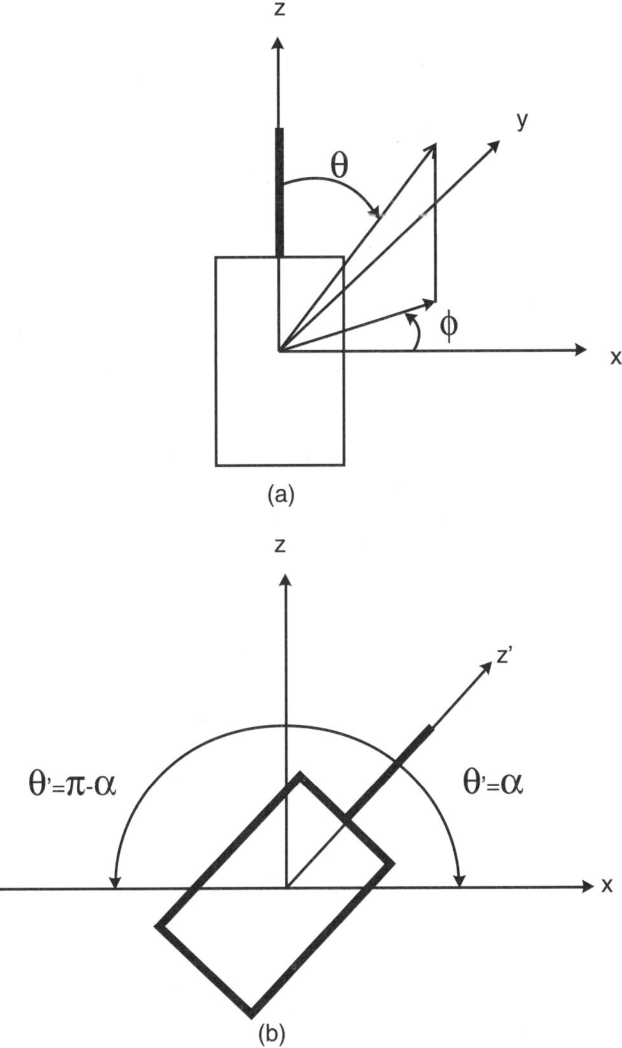

Figure 3.5 Inclination angles θ′, α of the (a) handset and (b) coordinate system.

Figure 3.6 shows an example of the average received electric field strength as a function of the angle of inclination, using the handset radiation pattern shown in Figure 3.2(a). The data sets S, M, and L represent small, medium, and large casing sizes, respectively.

Though the difference in the casing size is hardly noticeable when the inclination angle is less than 10°, it can be clearly seen that the antenna gain of even the smallest-sized casing is large for an inclination angle greater than

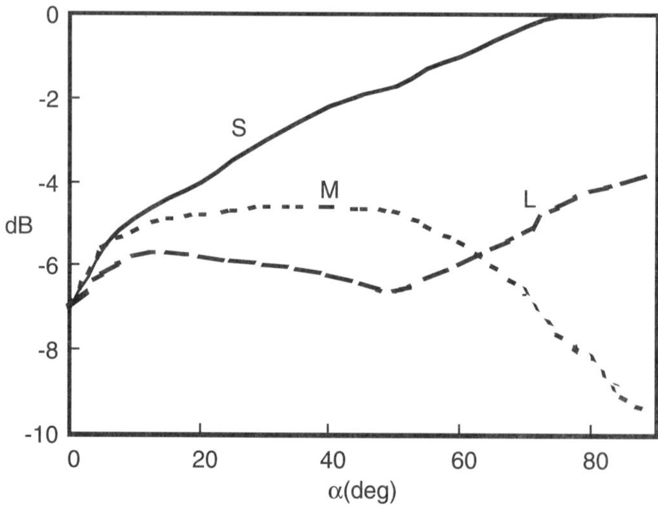

Figure 3.6 Inclination angle of handset and antenna effective gain. S, $L = 0.25\lambda$; M, $L = 0.65\lambda$; L, $L = 1.0\lambda$.

10°. We can conclude from these results that it is desirable to increase the effective gain of the antenna so that sidelobes do not appear in the radiation pattern.

3.1.3 Current Flow Measurements of the Handset Casing

Once the antenna has been installed onto the handset casing, simulations or measurements are required to find the current flow on the casing surface, but because it is difficult to include all of the electronic components that make up the handset in the simulation model, it is better to actually measure the current flow on the casing. Current flow can, however, be estimated indirectly by measuring the radiation pattern from the antenna plus the casing. This will then show exactly how the current is flowing on the casing surface.

For measurement of the radiation pattern, the influence of the feeder cables should be minimized. The current flowing on the cable is suppressed to a low level by covering it with ferrite beads and a balun. Alternatively, the feeder cable can be removed using a built-in miniature transmitter in the terminal. To measure the current distribution directly, a small loop antenna can be used as a magnetic flux detector to locate on which part of the casing the current is concentrated. The magnetic flux density B_z can be considered to be almost uniform inside the loop if the loop diameter is very small. The coordinate system of the loop is shown in Figure 3.7.

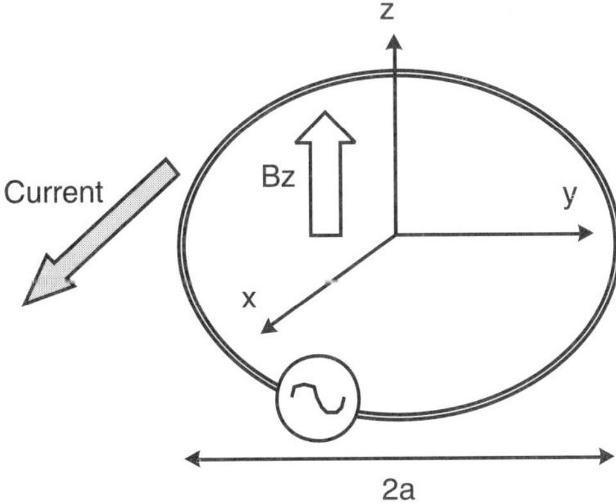

Figure 3.7 Small loop antenna and measured current.

From Faraday's law, when B_z changes with time the voltage generated across the gap in the loop is given by:

$$e = \frac{\partial \Phi}{\partial t} = -j\omega N \pi a^2 B_z \tag{3.3}$$

where Φ denotes the magnetic flux inside the loop, N is the number of loop turns, and a the loop radius. A small loop antenna is suitable as a probe for magnetic field detection because the induced voltage is proportional to the magnetic flux through the loop. The relationship between the magnetic field and the current flowing on the handset casing can be found using Ampere's law as shown in Figure 3.7.

The current flowing on the casing can be measured by scanning a small loop in the vicinity of the casing as shown in Figure 3.8. Because the current behavior on the casing is different for different antenna types, the most suitable antenna can be selected for the design of the handset antenna by measuring current distributions.

3.2 The Phantom: An Electrical Equivalent Model of the Human Body

Although the influence of the handset casing on the antenna operating characteristics is a major problem, the primary problem to be overcome is the

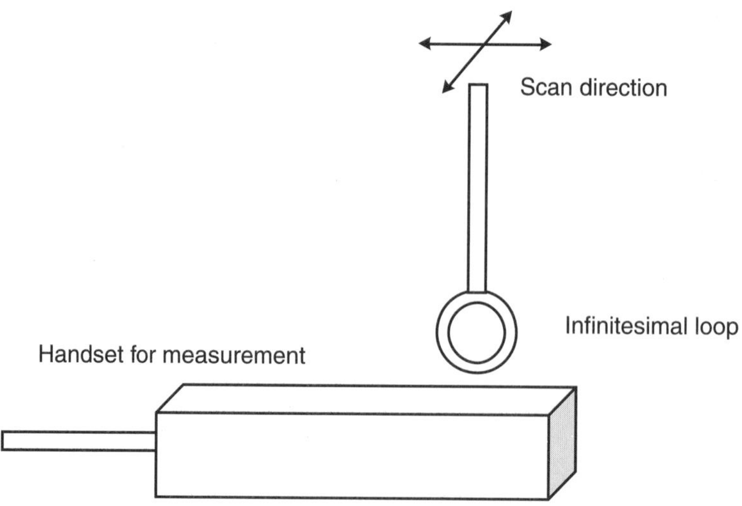

Figure 3.8 Measurement of current distribution on handset casing using a small loop antenna.

influence of the human body on the antenna. Because the handset is used in proximity to the head and body, it is strongly affected by these parts of the human body. Thus, for practical handset antenna designs the influences of the human body on the antenna must be accounted for as much as is possible. Moreover, even when under the influence of the human body, the antenna should still operate within specified limits dictated by the system designer.

The design of a handset antenna should also be such that the human body (specifically the head) is irradiated as little as possible. A medical technique that utilizes microwave radiation, called *diathermy,* is often used in the treatment of several different kinds of disease. This type of medical treatment was uncovered not long after the discovery of electromagnetic waves. However, since there was no theoretical basis for the analysis of this effect at that time, only the negative influences of electromagnetic waves on the human body were recognized.

Once it was realized that electromagnetic waves can heat human body tissues by various degrees, a safety standard was introduced indicating an acceptable degree of exposure of the human body to electromagnetic waves. The original safety standard defined the safe level of irradiation of the surface of the human body as 10 mW/cm^2 [3], but this was later lowered to 1 mW/cm^2 [4]. An application of this heating effect can be found in the treatment of cancer. It is known that cancer cells have a lower resistance to heat than normal body cells, thus irradiating both the cancer and normal cells with electromagnetic radiation can destroy the cancer cells. This type of medical treatment is referred to as *hyperthermia*. However, the purpose of this chapter is not to discuss the influence of electromagnetic waves on the human body but to evaluate whether or not the level of electromagnetic radiation radiated by the portable handset terminal satisfies the national safety standard.

With the cooperation of a volunteer, it is possible to measure the influence on the human body of electromagnetic radiation from the handset. In performing the experiment, one problem is that the volunteer must maintain the same posture as that adopted when the handset would be in actual use for long periods of time. Another problem is that the volunteer can be overexposed to electromagnetic radiation. There can also be considerable differences between individuals, thus making it difficult to achieve data repeatability.

The gain of a handset antenna was evaluated using the random field measurement method for 30 different test subjects and reported that the measured antenna gain had a variation of 6dB, depending on a difference in operators [5]. To obtain test data repeatability, the influence of the human body on the antenna can be evaluated using a special kind of mannequin called a *phantom*. The phantom is made from specific types of dielectric material that are electrically equivalent to the human body. In addition, the composition of the dielectric materials that make up the phantom (or the human body) can be entered into a computer in order to simulate the experimental test setup. The specifics of phantoms for numerical simulation and experimental usage are described in detail in the following sections.

3.2.1 Phantoms Used in Numerical Simulations

For mobile telephone development, it is necessary to evaluate the influence of the human head on the handset radiation characteristics. A European workgroup called Co-Operation Scientifique & Technique 244 (COST 244) has proposed two simplified models of the head for numerical evaluation purposes [6]. One model is a cube with a side length of 20 cm and the other is a sphere of radius 10 cm. Both the cube and the sphere are modeled on the basis that they are composed of a uniform dielectric material. An example of typically assumed material constants is given in Table 3.1. To model the skin, an outer shell of dielectric material of 5-mm thickness is added with a relative dielectric constant of 3 ($\varepsilon_r = 3$). A similar phantom described by the COST 244 work group was also prescribed in the United States by the IEEE Standards Coordinating Committees (SCC 34), which describes safety standards for irradiation by electromagnetic radiation. There is, however, a slight difference in the assumed material constants as shown in Table 3.1. Such a

Table 3.1
Electrical Parameters of the COST 244 (SCC 34) Phantom

f (MHz)	ε_r	σ
900	43 (42.5)*	0.83 (0.85)
1800	41 (41.0)	1.14 (1.65)

* Figures in parentheses are for SCC 34.

simple form of phantom is useful as a standard to check the validity of computation codes and measurement setups.

Because the composition of the human body is extremely complex, a phantom model that imitates the exact structure of the body (e.g., the head) is necessary for precise calculations.

3.2.2 Calculation Methods for Phantom and Handset Antenna Models

Numerical calculation methods such as the method of moments and the finite difference time-domain (FDTD) method are typical methods for the evaluation of such problems. The method of moments and other analytical techniques are only effective when the phantom is a symmetrical shape such as the cube or the sphere described in the previous section.

Analytical results are available for models such as the layered sphere phantom, since Green's function can be derived for such a case [7, 8]. Green's function is used for an electromagnetic field excited by a unit source vector under the given boundary conditions of the problem. Once the Green's function is derived, the electromagnetic fields are calculated for an arbitrary distributed source using that function.

The validity of numerical simulation methods such as the FDTD method, which is to be described later, has been validated by comparison with analytical models. To calculate the absorption of electromagnetic waves in thin layers such as the skin (as compared to the underlying structures of the body), a very large computer memory is required. Also, it has been found difficult to perform such calculations using the FDTD method. The layered structure model was therefore created to help alleviate some of these problems. An example of a sphere utilizing the layered structure model is shown in [9]. Figure 3.9 shows the model to be analyzed, and the corresponding parameters are given in Table 3.2. Using the layered-sphere model, it is thus possible to compute the absorption of the electromagnetic wave in a thin layer such as the skin. Table 3.2 shows the assumed material characteristics of the various layers in the model for the human head.

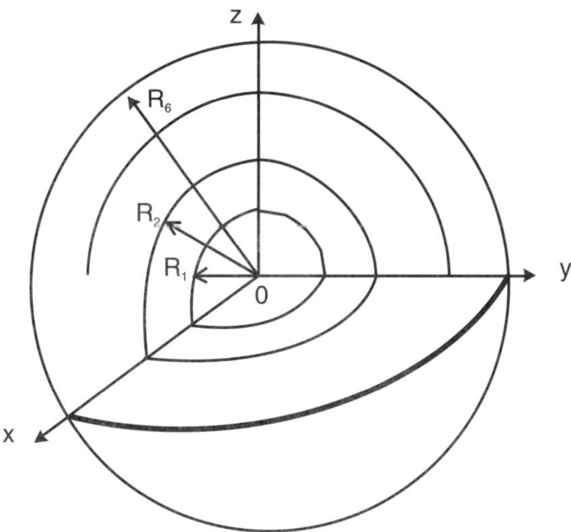

Figure 3.9 Layered-sphere phantom model.

Using the FDTD method it is possible to divide an object into very small volumes, referred to as *cells*. Once the real object has been subdivided into cells, the characteristics of the model can be calculated numerically, thereby giving a close approximation to the actual characteristics. For precise modeling, large computational resources are required to create the necessary small-sized cells, which then allow modeling of the handset structure and the operator's hand. Figure 3.10 shows an example of a phantom head model for numerical simulation purposes. The phantom used in numerical simulations is based on the anatomical chart of the human body [10]. A phantom with

Table 3.2
Parameters of Layered-Sphere Model (f = 2 GHz)

P	Layer	R_p (cm)	ε_r	σ (S/m)
6	Skin	a	47.5	1.33
5	Fat	a-0.15	6	0.10
4	Bone	a-0.27	5	0.20
3	Dura*	a-0.70	47.5	1.33
2	CSF*	a-0.80	83.2	1.33
1	Brain	a-1.10	59.4	1.00

* Dura = cranial dura matter; CFS = cerebrospinal fluid.

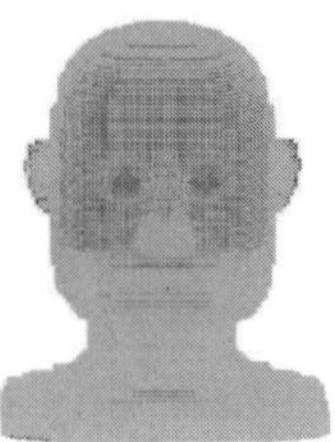

Figure 3.10 Head phantom model for calculations. (Courtesy of Prof. Itoh, Chiba University.)

a millimeter resolution accuracy can be made automatically from magnetic resonance images (MRI) and computer tomography (CT) data of the human body [11]. The phantom used for numerical simulations must be specifically tailored to the purpose of the calculations. Once a phantom for the simulation has been prepared, it is helpful to cross-reference the phantom cross-sectional composition with data recorded on the Internet home pages of the Visible Human Project [12] and the U.S. Federal Communications Commission (FCC). Both home pages contain data on the electrical characteristics of human body tissues versus frequency [13].

3.2.3 Phantoms for Use in Experimental Measurements

To evaluate the characteristics of a mobile handset operating near the human body as well as to confirm the validity of simulation models, experimental measurements using a physical model of a phantom are also required. Since cost and/or complexity preclude the construction of a phantom that is an exact replica of the human body, a phantom composed of two different kinds of uniform dielectric material is normally used. A plastic skin several millimeters thick, called the shell, is used to enclose a liquid and maintain the physical form of the phantom.

There are two main classifications of phantoms for experimental usage: (1) a *dry phantom,* which is made of a hard ceramic material, and (2) a *wet phantom,* which consists of a water solution or a jelly-like substance made into the desired shape. One problem is that the dry phantom tends to be very heavy although the hard material acts as a realistic imitation of the head, hand, and upper half of the body. However, the wet phantom offers the advantage that additives can be used to change the material characteristics of the phantom. Additionally, when performing electromagnetic wave irradiation experiments, it also has the advantage that an optional cut can be made through the body of the phantom to observe an arbitrary cut plane in the measurements.

On the other hand, a major disadvantage of wet phantoms when compared with dry phantoms is that, due to the type of material used in wet phantoms, it is difficult to preserve them for more than one month.

3.2.3.1 Dry Phantom

The human body consists of low-water-content structures such as the skin, fat, and bone, and high-water-content structures such as the brain, muscles, and internal organs. However, the electrical characteristics of the tissues vary greatly at frequencies of 10 MHz and less [14]. For the mobile communications operating frequency band of 800 MHz to 2 GHz, the dielectric-loss tangent is found to range from 0.1 to 3.0, and the relative dielectric constant from 20 to 70 [15]. For materials used in the construction of phantoms, it is not easy to obtain materials with a loss tangent (tan δ) of between 1 and 10. Ceramic materials, however, can be used to give a relative dielectric constant (ε_r) of between 10 and several 10,000s, in the microwave frequency band. A phantom with a loss tangent similar to that of the human body can be made by adding conducting powders to the ceramic [16]. The composition of the plastic shell used to form the phantom shape is Ba, Ca, Ti, Sn, and carbon powder, which then mimics the electrical characteristics of the living body. The electrical characteristics of the plastic and oxide powder are $\varepsilon_r = 3.1$, tan $\delta = 0.1$ at 1 MHz and $\varepsilon_r = 20$, tan $\delta = 0.02$ at 1 MHz, respectively. The dry phantom shown in Figure 3.11 proves particularly useful when the influence of the hand on the radiation characteristics of the handset has to be considered.

3.2.3.2 Wet Phantom

For the examination of electromagnetic irradiation effects on the human body, it is important to measure the temperature rise distribution inside the phantom and a dry phantom must therefore be cut into sections during the pro-

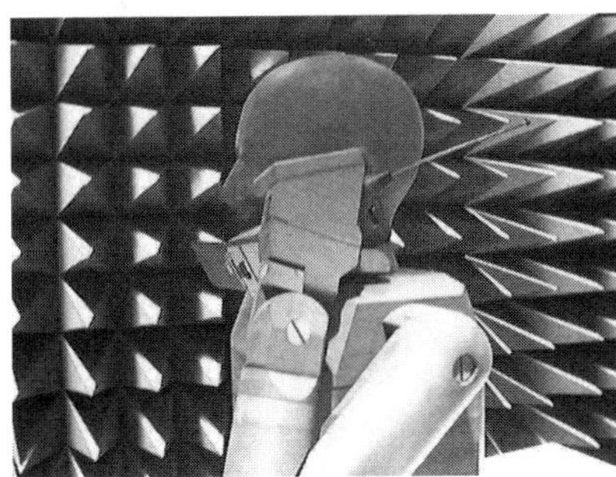

Figure 3.11 Ceramic dry phantom.

Figure 3.12 Wet phantom. (Courtesy of Prof. Itoh, Chiba University.)

duction process. The wet phantom is convenient, therefore, because an optional cut along one side is possible for making such measurements, as shown in Figure 3.12. For performing hyperthermia experiments, it is also possible to place the radiation applicator inside the wet phantom at a location of the user's choice.

In the construction of wet phantoms, a salt solution and agar have been the materials of choice for phantoms to be used in the microwave frequency band [17]. The loss tangent can be controlled by use of the electrolyte NaCl, although control of the relative dielectric constant has proven to be difficult. Because the material from which a wet phantom is made is equivalent to biological material, it also suffers the problem that it can decompose with time. For experimental usage, the best wet phantoms can be preserved for long periods of time as well as having suitable material characteristics. Such a phantom has been reported in [18]. If the food preservative dehydroacetic acid sodium salt (DASS) is added to the phantom material, preservation periods of 1 month or more are possible at normal temperatures. The phantom should also be covered by a thin film.

The electrical characteristics of the muscle tissue can also be imitated in the frequency range from 200 MHz to 2.5 GHz. An example of the types of material used in the composition of a wet phantom is shown in Table 3.3. In the 900-MHz frequency band, the relative dielectric constant can be controlled (35 to 65) using polyethylene powders, and the conductivity by introducing small quantities of NaCl (0.3 to 2.5). Figure 3.13 gives an example of actual data for wet phantom material composition.

3.2.3.3 Whole-Body Phantoms

In the 900-MHz operating band and higher, the hand and head are the dominant influences on mobile telephone radiation characteristics. For low-

Table 3.3
The Composition of the Phantom*

Ingredients	Muscle (g)	Brain (g)
Deionized water	3375	3375
Agar	104.6	104.6
Sodium chloride	39.2	23.1
DASS	2.0	2.0
TX-151	84.4	57.1
Polyethylene powder	337.5	548.1

* Phantom volume is about 3500 cm^3. Agar is for solidification, sodium chloride for conductivity control, DASS (dehydroacetic acid sodium salt) for preservation, TX-151 by Oil Center Research Inc., for gelling, and polyethylene powder for relative dielectric control.

frequency bands such as the 150-MHz band, which is used for multiple channel access (MCA) services, a whole-body phantom is required. This is because the wavelength of the radiation is considerably longer than at microwave frequencies and resonance effects are found to occur across the entire length of the human body. VHF band pagers are also used in proximity to the human body, thus it is also necessary to evaluate the influence of the human body on the operating characteristics of the pager.

A dry whole-body phantom can be made using only ceramics, whereas a wet whole-body phantom is made by filling a plastic mannequin with a salt solution and using a double cylinder made of vinyl chloride containing ma-

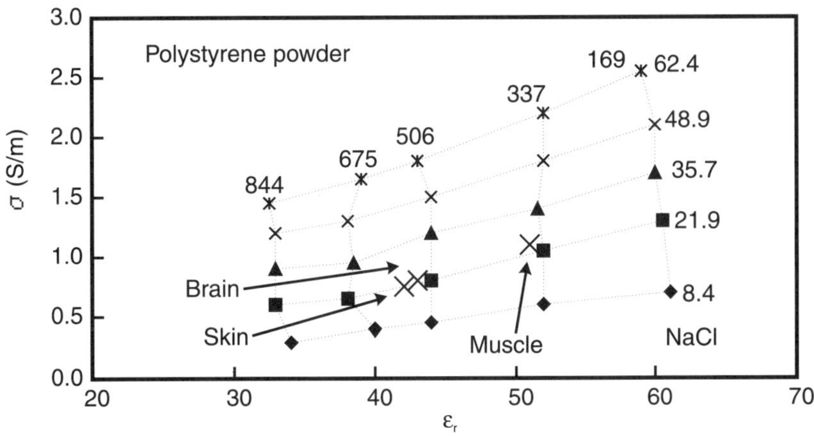

Figure 3.13 Phantom material and electrical parameters.

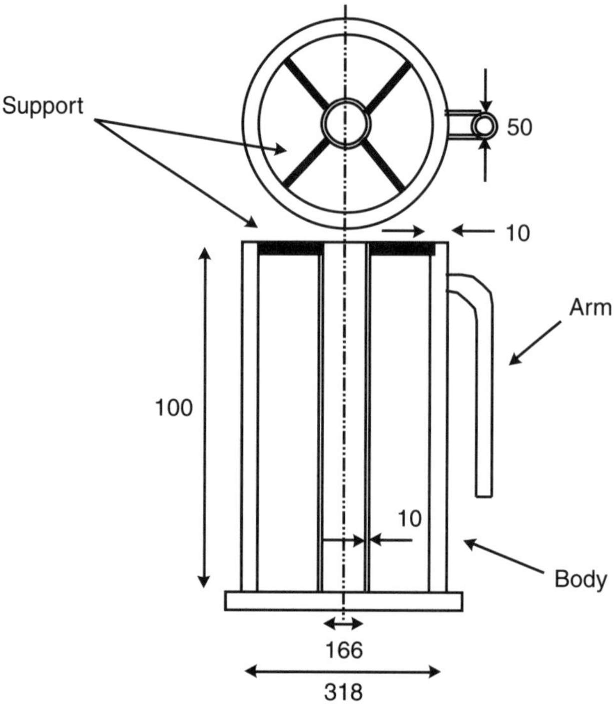

Figure 3.14 Double cylinder phantom (all units are in millimeters).

terial that is electrically equivalent to that of human muscle tissue. An example of such a whole-body phantom is shown in Figure 3.14 [15]. There is also a uniform phantom composed of rectangular parallelepipeds that occupies the same volume as a "standard" human body [18]. As an example, a phantom that is equivalent to the average Japanese man in his twenties is 40 cm × 16 cm × 166 cm, with a relative dielectric permittivity of $\varepsilon_r = 40.0$ and a conductivity of $\sigma = 0.32$ S/m.

3.3 Antenna Measurements Using a Phantom

This section presents measurement methods for evaluating the characteristics of a mobile handset antenna in proximity to various types of phantoms. Actual measured data of the antenna electrical and radiation characteristics are also presented. The measurement example presented shows the input impedance and radiation pattern of a standard dipole antenna placed near a sphere-shaped wet phantom and a dry phantom shaped like the upper half of the human body.

3.3.1 Measurements of Antenna Characteristics

The presence of the human body close to a handset antenna can change the input impedance characteristics and also distort the antenna free-space radiation pattern. The influences of the hand and the head on the antenna can be measured by fixing the antenna and handset to the hand of the dry phantom and introducing a coaxial feeder cable between the antenna feed point and the measurement equipment. As mentioned in Chapter 2, while performing the measurements, the position of the antenna (and cable) relative to the phantom should not be changed. A material such as styrene foam, with a relative dielectric constant close to 1 (almost equivalent to air), is used to hold the antenna in position during testing. Figure 3.15 shows the antenna measurement setup using a dry phantom.

For the antenna input impedance measurement, if the distance L between the feed point and the phantom surface is greater than a wavelength ($L > \lambda$), the measured results are unaffected by the presence of the phantom. Because the phantom acts as an obstacle that disturbs the radiation field from the antenna, the radiation pattern toward the phantom side will be changed.

Measurement examples are presented in a later section. By measuring the radiation efficiency of the antenna while it is in proximity to the phantom, the extent to which power is absorbed by phantom can be evaluated. As described in Chapter 2 for the measurement of the absolute radiation efficiency, by necessity, the measurement system becomes very large and the measurement procedure time-consuming.

For a relative measurement, the random field method is a well-known technique. This method was also explained in Chapter 2. Normally, for the random field measurement a human test subject is used to hold the terminal while the measurement is performed. However, the measurement error can be reduced by the use of a phantom. The mean effective gain (MEG) of

Figure 3.15 Antenna measurements using dry phantom.

the radiation pattern is measured inside a radio-frequency (RF) anechoic chamber and is used to evaluate the relative efficiency. In a real operating environment for a mobile communications system, an approximate antenna gain is calculated using the MEG in the horizontal plane, as defined in (3.1). As mentioned previously, this is because the signal arriving at the mobile terminal is coming from a point far away and the RF waves are thus concentrated in the horizontal plane. The relative gain of the antenna G_t is evaluated by taking the difference between the mean received field strength E_s, measured with a standard half-wavelength dipole antenna of gain G_s, and the measured value of the received field strength E_t taken with the antenna under test:

$$G_t = |E_t|^2 - |E_s|^2 + G_s \tag{3.4}$$

3.3.2 Measurement Examples Using a Sphere-Shaped Wet Phantom

Examples of measurements using a sphere-shaped wet phantom having a 10-cm radius and with a relative dielectric constant of $\varepsilon_r = 52 + j19$ are given in this section. The parameters to be fixed during the measurement procedure are (1) the direction of the antenna toward the phantom and (2) the distance between the phantom and the antenna.

For the measurements, a standard dipole antenna was used and two specific measurements were made: (1) measurement of the position of the maximum radiation from the antenna toward the phantom and (2) measurement of the position of the null in the radiation pattern, as shown in Figure 3.16. The measurements were made at a frequency of 2.5 GHz. The radiation patterns from the antenna in the E and H planes are shown in Figures 3.17(a) and (b), respectively. The pattern for only the E plane of Figure 3.16(b) is also shown in Figure 3.18, because this antenna position is axially symmetrical about the antenna. The H-plane pattern is omnidirectional for all the parameters.

We can see from Figures 3.17 and 3.18 that the radiation pattern on the phantom side ($180 \leq \theta \leq 360°$) is suppressed in Figure 3.17. The radiation strength toward the phantom becomes small because the input impedance characteristic of the antenna deteriorates, which will be mentioned later, if the antenna is too close to the phantom, as shown in Figure 3.17(a). The E-plane radiation profile (Figure 3.18) for the experimental setup shown in Figure 3.16(b) shows less change than that of Figure 3.16(a), because the null position of the radiation pattern faces toward the phantom.

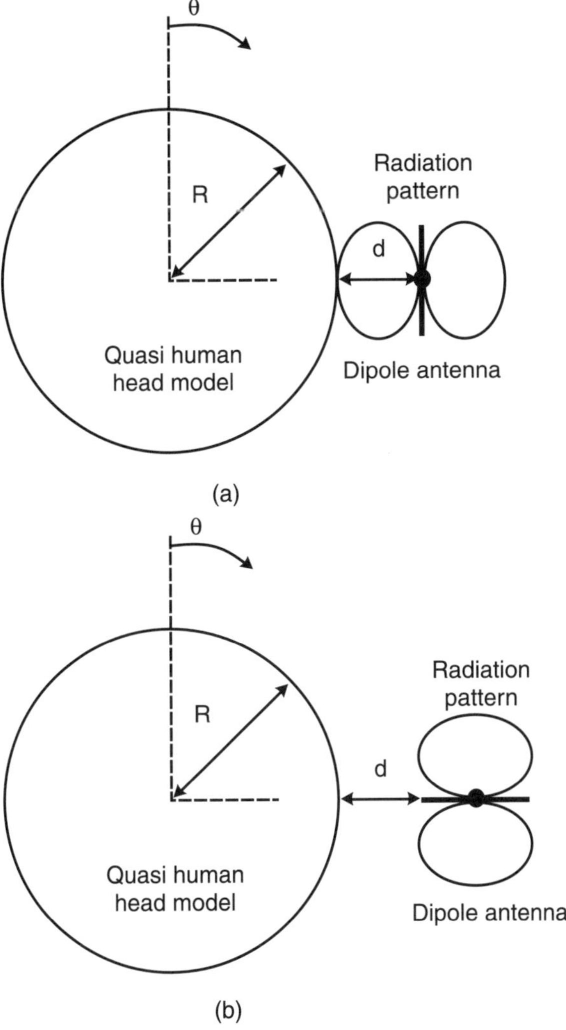

Figure 3.16 Antenna position with respect to the phantom: (a) the maximum radiation from the antenna toward the phantom and (b) the null in the radiation pattern.

The radiation pattern on the phantom side is caused by a creeping wave. The creeping wave is excited at around $\theta = 90°$, propagates along the phantom surface, and then radiates backward. The electromagnetic waves "creep" on the sphere surface, and this is the reason for the use of the term *creeping wave.*

The return loss characteristic of the antenna is also shown in Figure 3.19 for varying distances between the antenna and the phantom and also for both antenna orientations toward the phantom. We can see that the antenna

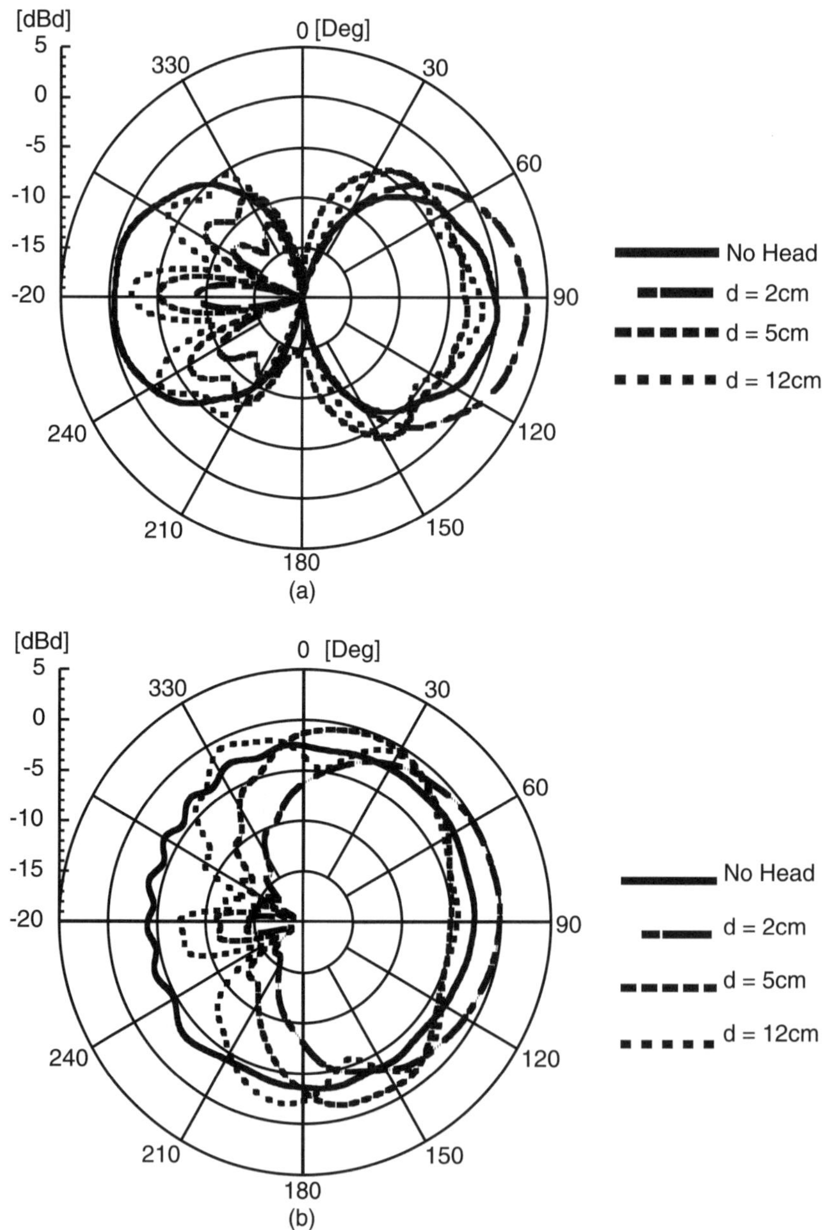

Figure 3.17 (a) E-plane and (b) H-plane radiation patterns for position (a) from Figure 3.16.

resonant frequency decreases, and the matching condition at the feed point deteriorates when the antenna is in proximity to the phantom, except for the curve for $d = 1$ cm in Figure 3.19(a). The resonant frequency for $d = 0$ in Figure 3.19(a), that is, with the antenna feed point attached to the phantom,

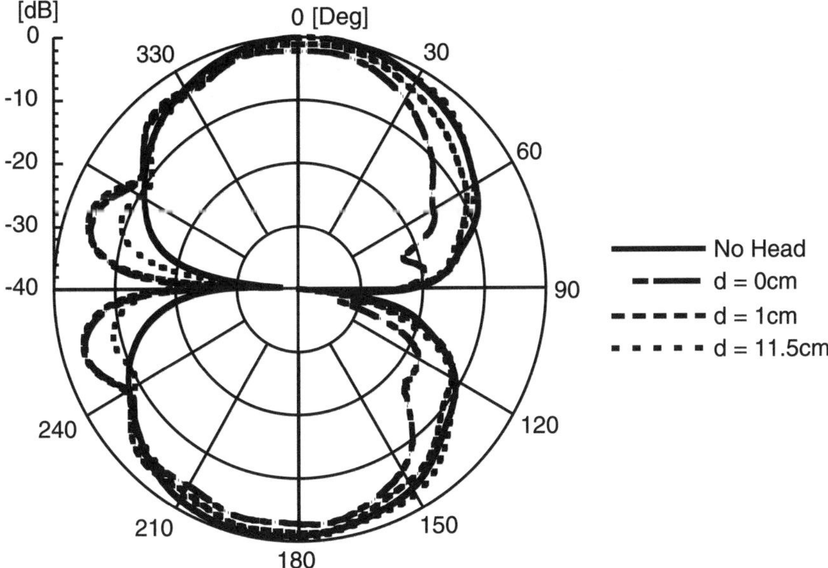

Figure 3.18 E-plane radiation pattern for position (b) in Figure 3.16.

goes to the high-frequency side. In this case, part of the antenna current flows to the phantom and results in a different tendency with respect to the other parameters.

If the antenna distance from the phantom is greater than one wavelength for the orientation shown in Figure 3.16(a), then the return loss characteristic is the same as if there were no phantom present. For an antenna orientation like that shown in Figure 3.16(b), the influence of the phantom can be neglected if the antenna is at a distance greater than a quarter wavelength.

3.3.3 Measurement Examples Using an Upper Body Model Dry Phantom

This section presents a measurement example using a dry phantom shaped like the upper half of the human body. Figure 3.20 shows the experimental setup used to examine three different antenna orientations. This is the same experimental setup as that discussed in the preceding section.

As with the previous experiment, when the distance between the antenna and the phantom is more than one wavelength, the influence of the phantom on the antenna can be neglected. As shown in Figures 3.21(a) and (b), the main radiation component toward the phantom ($0 \le \theta \le 180°$) is suppressed, as is the radiation in the opposite direction ($180 \le \theta \le 360°$).

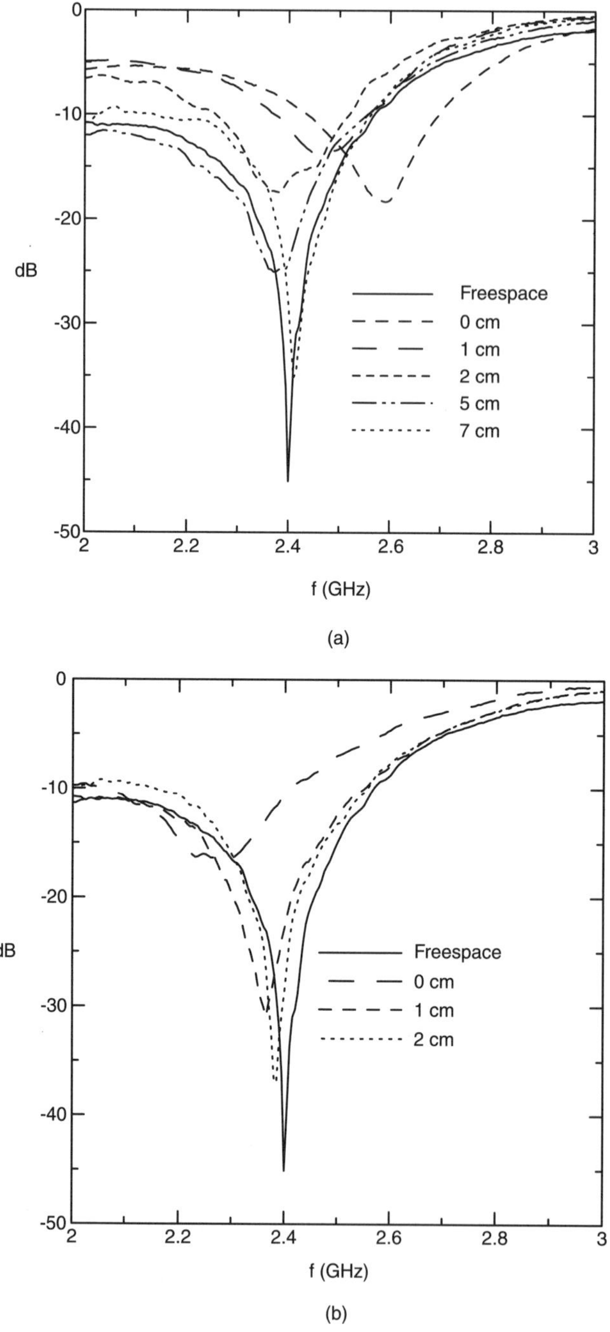

Figure 3.19 Dependence of input characteristics on the relative position of antenna and phantom for varying antenna positions. (a) Antenna orientation shown in Figure 3.16(a), and (b) antenna orientation shown in Figure 3.16(b).

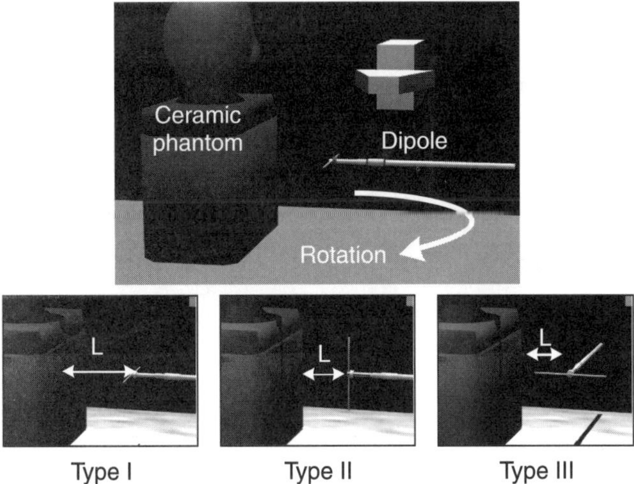

Figure 3.20 Antenna position toward phantom.

The main radiation component is E_θ for type I and E_ϕ for type II. This suppression is caused by the impedance mismatch at the antenna feed point. A unique phenomenon, the radiation of the cross-polarization component to the phantom side, can be seen in Figure 3.21(c). The creeping wave is strongly excited in this condition.

The variation of the MEG in the horizontal plane versus distance from the phantom is shown in Figure 3.22. It can be assumed that the MEG does not change when the distance between the antenna and the phantom is greater than one wavelength. The radiation pattern on the opposite direction from the phantom spreads out evenly in the H plane, while the E-plane radiation profile contains ripples. These ripples are caused by the influence of the upper half of the body and are still evident even when the distance between the phantom and the antenna becomes more than wavelength.

3.4 SAR Measurement Using a Phantom

This section presents examples of the regulation standard of specific absorption ratio (SAR) as a safety standard for electromagnetic wave irradiation of the human body. The detailed definition of SAR and how to measure it using a phantom are also presented together with an example of electric field strength distribution measurement inside a sphere-shaped wet phantom.

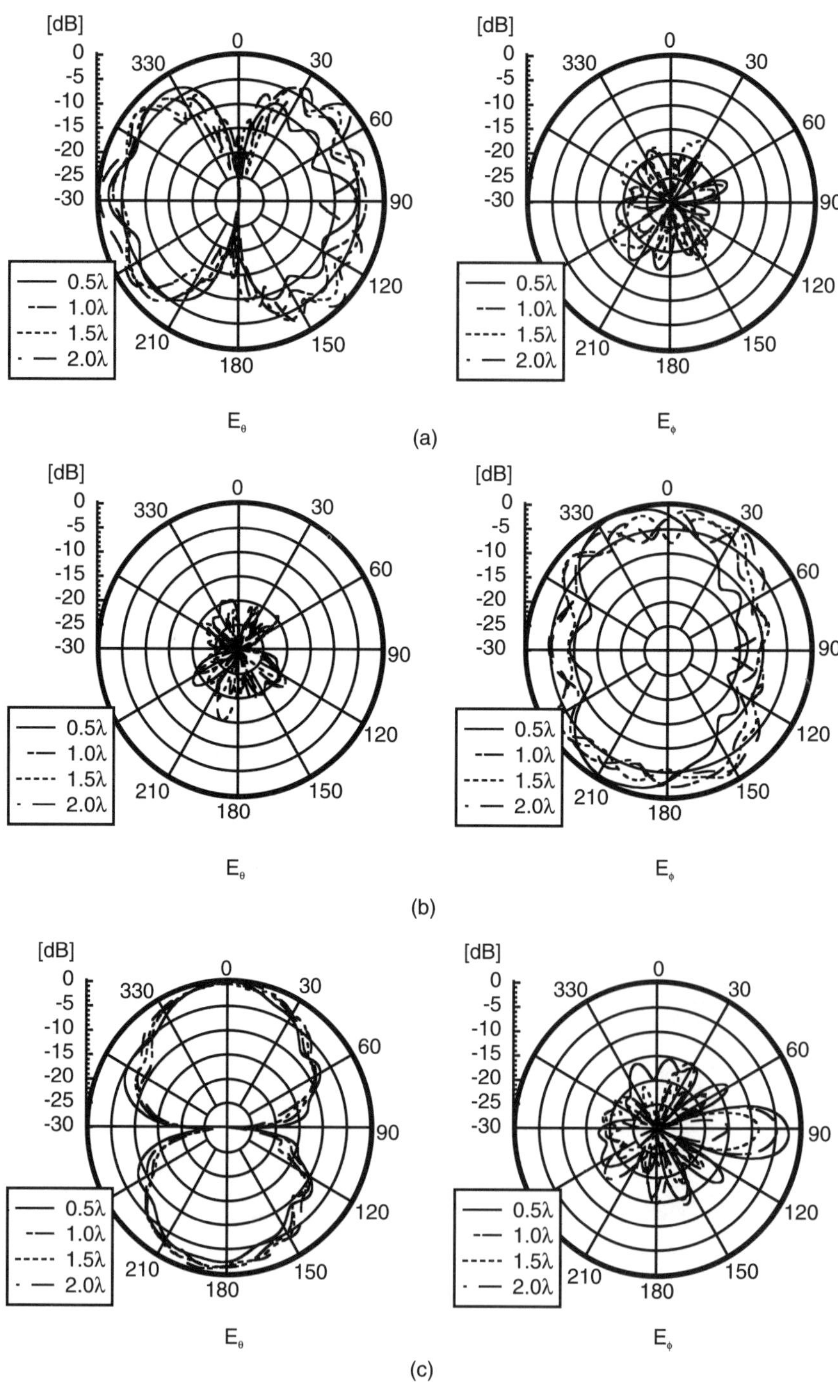

Figure 3.21 Radiation pattern of dipole antenna near dry phantom: (a) type I, (b) type II, and (c) type III.

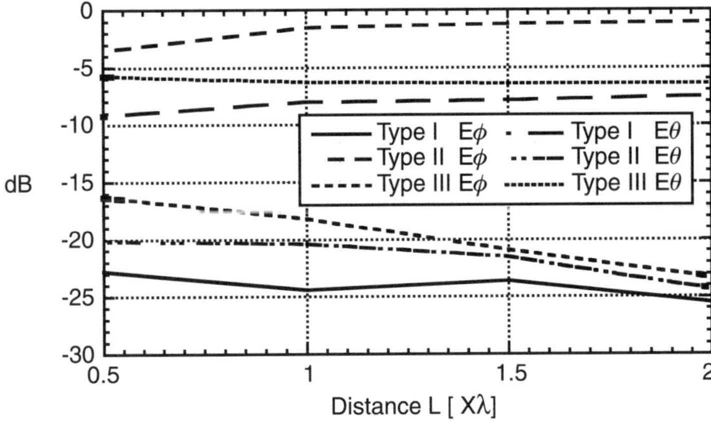

Figure 3.22 MEG for dipole antenna located near dry phantom.

3.4.1 Standard Value of SAR

The SAR (in watts per kilogram) is used as a safety standard for the irradiation level of electromagnetic waves applied to the living body. The SAR is a measure of the heat energy absorption in unit time by a living body. Three definitions of the SAR are classified: the absorption ratio by short pulse waves, the whole-body average SAR, and the localized SAR. The latter two are defined as a mean value during a given time. When the temperature rise is more than 1°C due to the heat source from the outside, it is assumed that there is some influence on the living body. The corresponding whole-body average SAR is then 2 to 3 W/kg [19].

The localized SAR is used mainly as the standard of irradiation level for portable terminals. On the other hand, an incident power level at the body surface is often used for the safety standard in addition to the SAR. The standard example is 1 mW/cm^2. The SAR safety standards used in Japan, the United States, and Europe are shown in Table 3.4.

Table 3.4
SAR Safety Standards for Japan, the United States, and Europe

Region	SAR (mW/kg)	Mass to Average (g)	Average Time (min)
Japan	8.0	1	6
United States	1.6	10	30
Europe	2.0	1	6

3.4.2 Definition of SAR and the Measuring Method

To evaluate the influence of electromagnetic (EM) wave irradiation on the living body, the SAR is defined as an amount of EM energy absorption in a unit mass as follows:

$$\text{SAR} = \frac{\sigma E^2}{\rho} \tag{3.5}$$

where the conductivity, the effective electric field amplitude, and the material density are denoted by σ (S/m), E (V/m), and ρ (kg/m^3), respectively. Therefore, the SAR at the measurement point is obtained from the electric field distribution inside. The phantom should be allowed to warm up for a period of more than 30 minutes in the measurement environment, as shown in Figure 3.23, so that the temperature distribution inside the phantom can become uniform. After that, the phantom is irradiated by the antenna installed in the neighborhood of the phantom for 10 to 100 seconds.

A standard dipole antenna is widely used in this type of measurement. The irradiation time used is the time required to increase the temperature of the phantom by more than 1°C. After irradiation, pictures are taken of the observation plane of the phantom with thermography, to give the two-dimensional distribution of temperature rise [20]. Figure 3.24 shows the measurement example using a sphere-shaped phantom with the electrical parameters of COST 244. In this case, the measurement frequency was 900 MHz. This type of measurement is called a *split phantom method*, because the observation plane of the phantom is cut in advance so that the temperature inside of the phantom can be measured very quickly after irradiation.

Because the SAR is based on a temperature rise in unit time, the heat diffusion from the sample during the measurement time can be neglected.

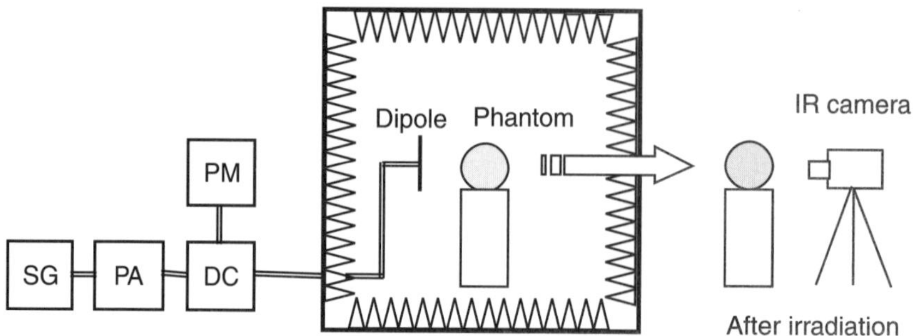

Figure 3.23 SAR measurement setup using phantom.

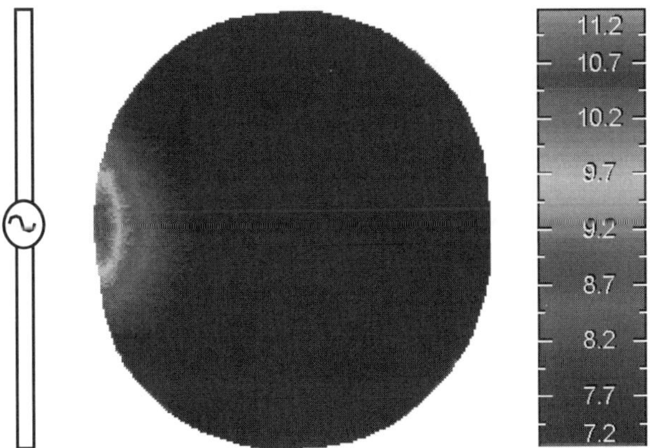

Figure 3.24 Temperature distribution inside phantom measured by thermography. (Courtesy of Prof. Takahashi, Musashi Institute of Technology.)

The SAR at any time is defined by the specific heat c (J/kg K) and the temperature rise ΔT (°C) at the observation point of the sample during the measurement time Δt:

$$\mathrm{SAR} = c\frac{\Delta T}{\Delta t} \tag{3.6}$$

For the SAR measurement using this definition, the electromagnetic waves are generated with the antenna installed either inside of or outside the liquid phantom, and the temperature rise inside the liquid can be measured directly with the thermometer [21].

3.5 Measurements Using a Human Body

The advantages of measurements using a phantom is that the characteristics thus obtained have good repeatability. However, the antenna characteristics are significantly changed by the way in which the handset is held and the inclination angle of the handset, which are different for each operator. Measurements using a real human body as an operator are necessary to confirm the validity of the experimental results with the phantom and the application range of the phantom measurements. A measurement example using a human operator is given in this section using the random field measurement and the mean effective gain of the radiation pattern.

3.5.1 Effect of the Operator's Hand

The antenna characteristics are affected by the hand and the head to a much greater extent than by the casing size and the antenna type used in the design of the handset antenna. Therefore, this section first discusses the influence of the hand. The change in the input characteristics of the handset due to the operator's hand is measured via the cable connected to the antenna feed point. Ferrite beads are mounted on the coaxial cable near the feed point in order to suppress the leakage current flowing on the surface of the cable.

In this return loss measurement, the operator's hand absorbs part of the current flowing on the casing, which then reduces the resonance frequency and expands the frequency bandwidth, as shown in Figure 3.25. It is therefore important to suppress the current flowing on the casing in order to reduce the influence of the hand. In practice, for handsets in commercial use, any such change in the input impedance when the operator holds the handset has to be within the range of the specification. This is an essential feature of handset design.

The effect of the hand is more serious for designs with built-in antennas, because the top surface of the built-in antenna is often covered by the hand when the handset is in use. When a part of the antenna is covered by the hand, the input impedance characteristics of the built-in antenna installed

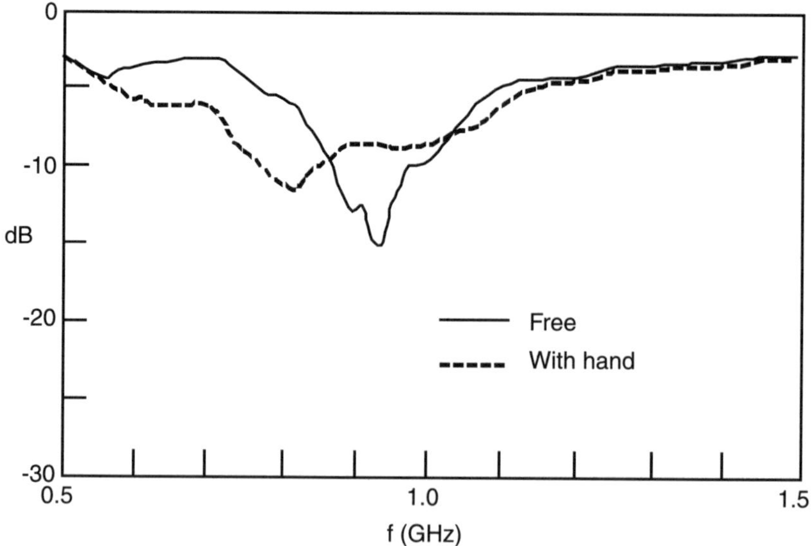

Figure 3.25 Change in input characteristics of handset antenna due to hand.

in the rear of the casing change greatly [22]. In designs with built-in antennas, the antenna location should be chosen so that it is not covered by the hand.

3.5.2 Loss in Antenna Gain Due to the Human Body

This section considers the loss in antenna gain of the handset terminal caused by the human operator, when measured by the mean effective antenna gain method. The operator using the handset terminal stands on a large rotating table inside the anechoic chamber, and the radiation pattern is measured. The loss due to the operator is obtained by normalizing the measurement value of the standard dipole antenna. The results are based on an average of measurements taken with seven adults [23]. The loss caused by the human body with a monopole antenna is about 3 dB higher than that with a sleeve antenna in the 900-MHz band. This is because the effective antenna gain depends on the length of antenna that protrudes beyond the head. The same measurements were carried out in a four-door sedan-type vehicle. In this case, the handset was held by the operator inside the vehicle in an open measurement site. The loss caused by the vehicle was approximately 9 dB, with the handset held in the left hand on the left side driver's seat.

3.5.3 Random Field Measurements of Antenna Gain

Good repeatable data are obtained using the MEG method in an environment that is electrically stable such as an anechoic chamber or an open site. However, because the position in which a handset terminal is held is not precisely the same for each operator, measurement errors are observed in the effective antenna gain. These errors can be eliminated by fixing the handset to the operator using a supporting structure made from foam material. Although this supporting structure is effective in decreasing measurement errors, it does not represent real usage of the handset terminal. The observed errors should be considered in the antenna design of the handset. In other words, these errors can be considered as a margin in the system design. To find the range of this margin, the errors in antenna gain are measured for several operators in conditions of actual use. The measurement of the antenna gain under these conditions is carried using the random field measurement method explained in Section 2.3.4. Such measurements can show the influence on antenna gain of different ways of using the handset and different ways in which it can be held by the operator.

Examples of measurements made using the random field measurement method are shown first. These are relative measurements using commercially available handsets and the measured data are not normalized with those from

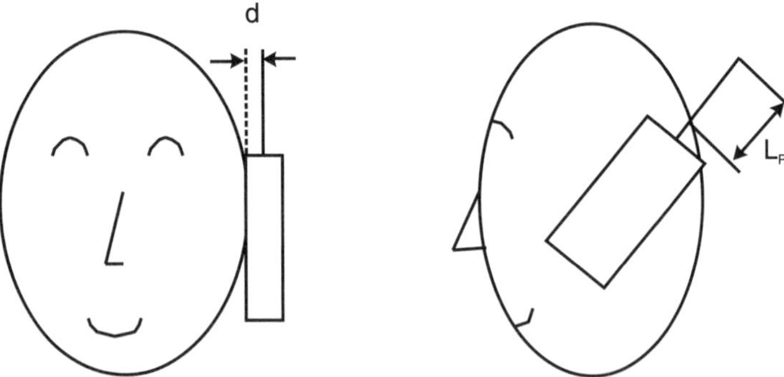

Figure 3.26 Antenna position for head.

a standard antenna. The measurement parameters relate to the antenna position with respect to the head; L_p is the length of antenna protruding beyond the head, and d is the spacing between the head surface and antenna. These parameters are defined with the handset antenna in the extended position. Figure 3.26 shows which values of L_p and d were used. Measured data shown in Figure 3.27 are the level of the uplink from the handset terminal

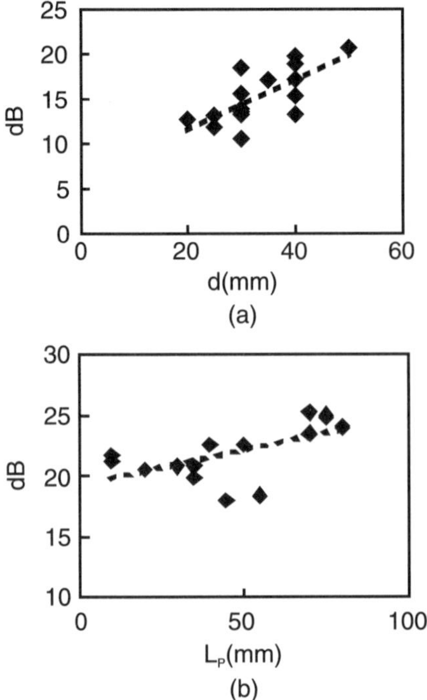

Figure 3.27 Effect of position of head and antenna on relative antenna gain with (a) d and (b) L_p.

picked up at the base station, with a measurement frequency of 900 MHz. Only one operator used the handset throughout the whole series of experiments. The uplink level is approximately proportional to the length L_p and the distance d. These results show that the effective gain of the antenna can be increased if the antenna is positioned away from the head. The antenna position giving a large length L_p in the extended position is also effective in increasing the antenna gain.

3.5.4 Gain Measurements on a Handset Antenna Using the Random Field Measurement Method

This section presents information on the deviations in antenna gain caused by the operator's hand during random field measurements. In addition, measurements with a tilted handset are also presented [5]. The handset is used near the head with a tilt angle similar to that in of a real-world situation. The three parameters that change the antenna gain are the hand, the head, and the inclination angle of the handset.

Figure 3.28 shows the antenna gain for four operators, where the measured data are normalized with respect to reference data in terms of the receiving electric field strength of a standard dipole antenna. In this measurement, each operator holds the same terminal near the head with the slant angle of 60° from the vertical direction. The result is obtained by taking the mean of the data from five separate measurements for each operator.

Figure 3.28 also shows the measurement result with the operator's arm stretched out straight in front of the body in order to separate the effect of

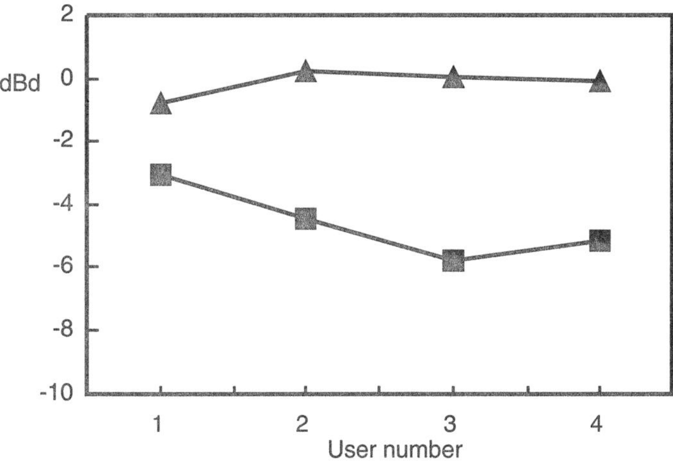

Figure 3.28 Effect of hand on relative antenna gain.

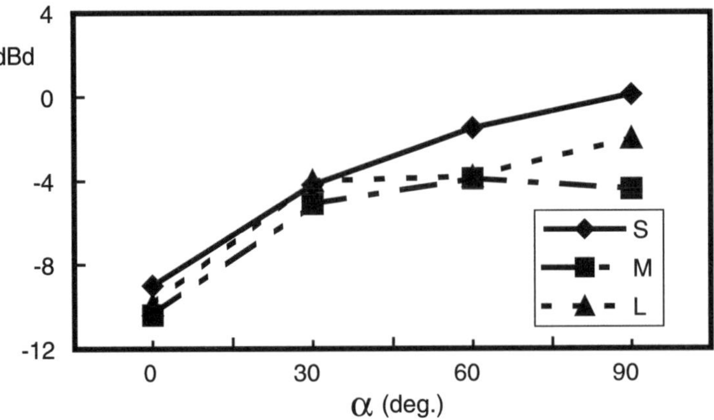

Figure 3.29 Effect of tilt angle on relative antenna gain.

the hand and of the head. When the operator's arm is outstretched and the handset terminal is held in the vertical position, the measured data correspond closely with those of the standard dipole antenna used as a reference. This result indicates that the antenna gain is not affected by the operator's hand. The terminal used for this measurement has electrical characteristics that result in a very small current flowing on the casing. If a large difference is obtained in this kind of measurement, the indication is that some current is flowing on the casing. This is another check method to find the effect of the current on the casing. The operator's body does not change the antenna gain and the distance between the handset and the body is about 50 to 60 cm. This distance is greater than a wavelength, and the input characteristics of the antenna are not affected. Because the phantom measurement gives the same results, a distance that is more than one wavelength is sufficient to allow the effect of the operator's body to be neglected.

When the handset terminal is used in a slanted position, the antenna gain is calculated using the MEG and the radiation pattern in the horizontal plane, as shown in Section 3.1. The antenna gain of a tilted handset, measured by the random field method, is also presented here to compare two results from different measurement methods. Figure 3.29 shows the measured gain of the slanted antenna using the random field measurement method. The handset parameters are the same as those of Figure 3.6. Here, the handset terminal under test is fixed to a wooden pole on the cart, at a height of 1.5m above the ground. Because the results of Figure 3.29 are the same as those using the MEG, as shown in Figure 3.6, the MEG method can be used for the evaluation of the radiation pattern of the handset terminal.

References

[1] Hirasawa, K., and M. Haneishi, *Analysis, Design, and Measurement of Small and Low-Profile Antennas*, Norwood, MA: Artech House, 1992, Chap. 1.

[2] Sekine, S., and T. Maeda, "The Radiation Pattern Characteristics of a λ/4-Monopole Antenna Mounted on a Conducting Body With a Notch," *Proc. IEEE AP/S Symp.* 1992, pp. 65–69.

[3] ANSI-66.

[4] "American National Standard Safety Levels With Respect to Human Exposure to Radio Frequency Electromagnetic Fields, 3 kHz to 300 GHz," IEEE C.95.1-1991, 1992.

[5] Arai, H., N. Igi, and H. Hanaoka, "Antenna-Gain Measurement of Handheld Terminals at 900 MHz," *IEEE Trans. Vehicular Technology*, Vol. 46, No. 3, Aug. 1997, pp. 537–543.

[6] d'Inzeo, G., "Proposal for Numerical Canonical Models in Mobile Communications," *Proc. COST 244 Meeting on Reference Models for Bioelectromagnetic Test of Mobile Communication Systems,* Nov. 1994, pp. 1–7.

[7] Hizal, A. H., and Y. K. Baykal, "Heat Potential Distribution in an Inhomogeneous Spherical Method of a Cranial Structure Exposed to Microwave Due to Loop or Dipole Antenna," *IEEE Trans. Microwave Theory and Techniques*, Vol. MTT-26, 1978, pp. 607–612.

[8] Amemiya, Y., and S. Uebayashi, "The Distribution of Absorbed Power Inside a Sphere Simulating Human Head in the Near Field of a λ/2 Dipole Antenna," *Trans. IECE Japan*, J66-B, 9, 1982, pp. 1115–1122.

[9] Jones, W. T., and R. J. Spiegel, "Resonance Absorption of Microwave by the Human Skull," *IEEE Trans. Biomed. Engineering*, BME-22, Vol. 6, 1975, pp. 457–463.

[10] Sullivan, M., O. P. Gandhi, and A. Taflove, "Use of the Finite-Difference Time-Domain Method for Calculating EM Absorption in Man Models," *IEEE Trans. Biomed. Engineering*, Vol. 35, No. 3, 1988, pp. 179–185.

[11] Hombach, V., et al., "The Dependence of EM Energy Absorption Upon Human Head Modeling at 900 MHz," *IEEE Trans. Microwave Theory and Techniques*, Vol. MTT-44, No. 10, 1996, pp. 1865–1873.

[12] http://www.nlm.nih.gov/research/visible/visible_human.html.

[13] http://www.fcc.gov/fcc-bin/dielec.sh.

[14] Cook, H. F., "The Dielectric Behavior of Some Types of Human Tissues at Microwave Frequencies," *British J. Applied Physics*, Vol. 2, 1951, pp. 292–300.

[15] Itoh, K., "Human Models for the Research of Antennas and Propagation," *J. IEICE*, Vol. J82, No. 9, 1999, pp. 956–966.

[16] Kobayashi, T., et al., "Dry Phantom Composed of Ceramics and Its Application to

SAR Estimation," *IEEE Trans. Microwave Theory and Techniques*, Vol. MTT-41, No. 1, 1993, pp. 136–140.

[17] Chan, K. W., "Microwave Phantoms," *Proc. COST 244 Meeting on Reference Models for Bioelectromagnetic Test of Mobile Communication System,* Nov. 1994, pp. 27–31.

[18] Itoh, K., et al., "Development and the Characteristics of a Biological Tissue-Equivalent Phantom for Microwaves," *Trans. IEICE,* Japan, B-II, Vol. J81-B-II, No. 12, 1998, pp. 1126–1135 (in Japanese).

[19] Sauders, R. D., et al., "Biological Effects of Exposure to Non-Ionizing Electromagnetic Fields and Radiation, III. Radio Frequency and Microwave Radiation," Nat. Radiol. Prot. Board Report NRPB-R240, 1991.

[20] Guy, W., C. Chou, and B. Neuhaus, "Average SAR and SAR Distributions in Man Exposed to 450-MHz Radio Frequency Radiation," *IEEE Trans. Microwave Theory and Techniques*, Vol. MTT-32, No. 8, 1984, pp. 752–762.

[21] Kaouk, Z., et al., "A Finite Element Model of a Microwave Catheter for Cardiac Ablation," *IEEE Trans. Microwave Theory and Techniques*, Vol. MTT-44, No. 10, 1996, pp. 1848–1854.

[22] Jensen, M. A., and Y. Rahmat-Samii, "EM Interaction of Handset Antennas and a Human in Personal Communications," *Proc. IEEE*, Vol. 83, No. 1, 1995, pp. 7–17.

[23] Hill, C., and T. Kneisel, "Portable Radio Antenna Performance in the 150, 450, 800, and 900 MHz Bands 'Outside' and In-Vehicle," *IEEE Trans. Vehicular Technology*, Vol. 40, 1991, pp. 750–756.

4

Base Station Antenna Siting, Measurement, and Maintenance

This chapter starts by presenting the design procedure used for siting base station antennas in cellular systems. The site location is based on statistical data of the propagation characteristics, which are summarized as a set of compact equations derived from these data, and the procedure is called the Okumura-Hata estimation method. The chapter describes this estimation method in detail, together with another method that is appropriate for the design of small size cells. Following site location, the base station antenna is then designed to satisfy the specifications. The design procedure for a cellular system is the second topic in the chapter.

At the development stage, the designed base station antenna characteristics are evaluated by making use of prototype models. The measurement method, particularly for cellular base station antennas, is the third topic described in the chapter. A unique measurement performed on base station antennas is the high-power test. This chapter describes the special features associated with the high-power treatment of the antenna and its measurement method in detail. The last topic in the chapter is the evaluation of the performance of the base station antenna. Tests needed before product shipping include electrical, mechanical, and reliability performance checks. These tests are described in detail together with practical examples.

4.1 Base Station Antenna Siting

The location of the base station is the most important issue for cellular systems. The coverage area of a cellular base station ranges from a radius of

0.5 to 20 km, and is determined by the number of expected users inside the coverage area and the geographical features in the locality. The Okumura-Hata method is widely used to estimate the propagation loss from a base station with a cell radius between 1 and 20 km. The Okumura-Hata estimation method described in this section has been used recently in microcellular systems with a cell radius below 0.5 km, in order to increase the user capacity of a single base station. The propagation loss estimation method is also presented for a cell of a similar small size. The selection of base station location is based on these propagation loss estimations.

4.1.1 Design of Macrocell Siting Using the Okumura-Hata Curve

In mobile communication systems, the propagation characteristics are predicted by an estimation method based on large quantities of data obtained from propagation loss experiments, described in Section 1.2.2. Geographical features inside a range of several tens of kilometers are roughly divided into a number of different types, and the propagation loss is estimated for these different types. This technique is well known as Okumura's curve. The variation of the propagation loss evaluated from measurement data at 150, 450, 900, and 1500 MHz defines the original Okumura curve. This section explains how this method can be used to estimate the propagation loss, described in Section 1.1 [1].

Geographical features can be divided roughly into irregular and semiflat. The irregular features are classified into hills, isolated mountains, inclined geographical features, and so on. But it is within semiflat geographical features that land mobile communication services are mainly used. Within semiflat geographical features, an undulating plane is defined as having a height of less than 20m. Local geographical features and buildings are described by a compensation value based on a place where the undulations are small. Such semiflat geographical features are classified as being in the city, in the suburbs, and in open areas. An interpretation for a city area, where the buildings have more than two stories, is that the proportion of buildings is equivalent to 23% of the area [2]. Figure 4.1 shows an example in the 900-MHz band for a city area. Distances up to 10 km from the base station are shown to a log scale, whereas greater distances are shown to a linear scale; this is a unique feature of Okumura's curve.

It is convenient to calculate data using the closed-form equation, although propagation loss can be read directly from the Okumura curve. The approximate equation of the Okumura curve, as derived by Hata, is shown below [3].

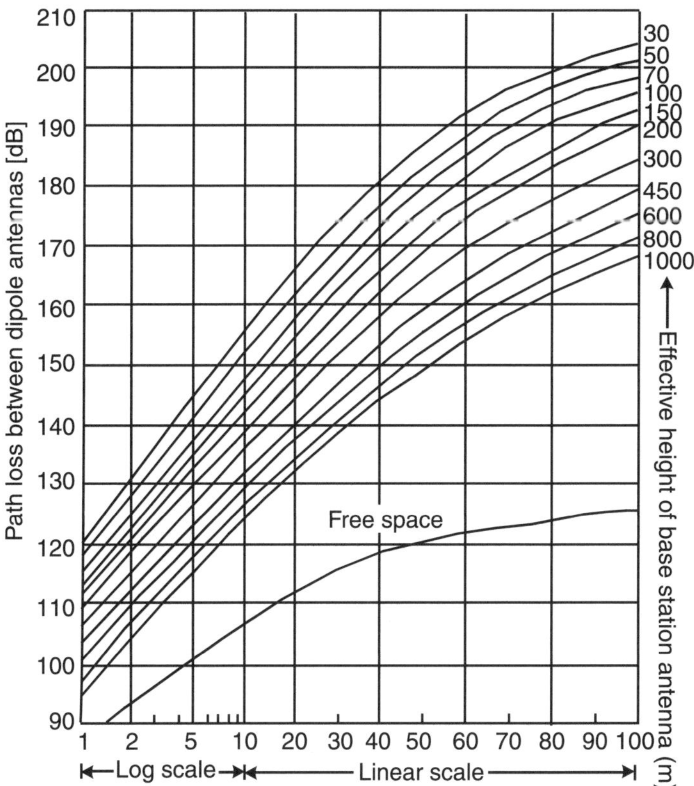

Figure 4.1 Okumura's curve for city area in 900-MHz band.

The propagation loss L_P in a city area is expressed by an equation at a frequency f MHz (150 to 1500 MHz), for a mobile station antenna of height h_m m (1 to 10m), with a distance between base station and mobile station of r m (1 to 20 km) as follows:

$$L_p = 69.55 + 26.16 \log_{10} f - 13.82 \log_{10} h_b - a(h_m) + (44.9 - 6.55 \log_{10} h_b) \log_{10} r \quad \text{dB} \tag{4.1}$$

where the antenna effective height $a(h_m)$ is defined as the compensated antenna height of the mobile terminal. For medium and small cities:

$$a(h_m) = (1.1 \log_{10} f - 0.7)h_m - (1.56 \log_{10} f - 0.8) \tag{4.2}$$

For large cities:

$$a(h_m) = \begin{cases} 8.29(\log_{10} 1.54 h_m)^2 - 1.1 & \text{for } f \leq 200 \text{ MHz} \quad (4.3a) \\ 3.2(\log_{10} 11.75 h_m)^2 - 4.97 & \text{for } f \geq 400 \text{ MHz} \quad (4.3b) \end{cases}$$

As shown in Figure 4.2, the antenna effective height is approximated by the height of the average surface of the Earth at a point where the mobile terminal is used at between 3 and 15 km from the base station antenna.

Next, the propagation loss is considered in areas that can be classified as suburban or pastoral and these include such areas as those along a road where trees and houses disturb the propagation of electromagnetic waves. Under these conditions:

$$L_p = L_p\{\text{city}\} - 2\{\log_{10}(f/28)\}^2 - 5.4 \quad \text{dB} \tag{4.4}$$

In open areas where high trees, buildings, and other obstructions do not exist within a range of 300 to 400m of the mobile station in the arrival direction, the propagation loss is given as:

$$L_{ps} = L_p\{\text{city}\} - 4.78\{\log_{10} f\}^2 + 18.33 \log_{10} f - 40.94 \quad \text{dB} \tag{4.5}$$

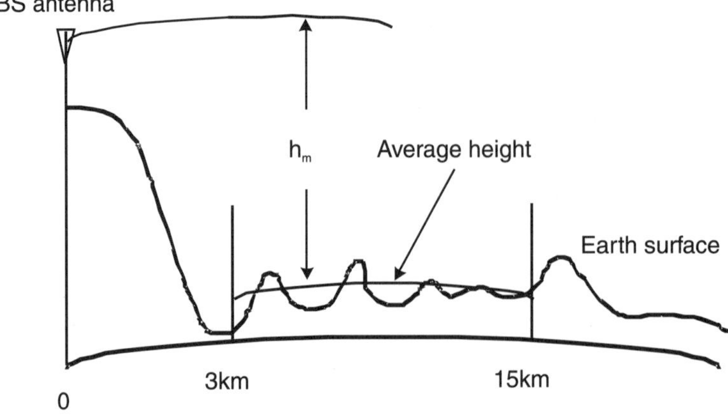

Figure 4.2 Definition of antenna effective height.

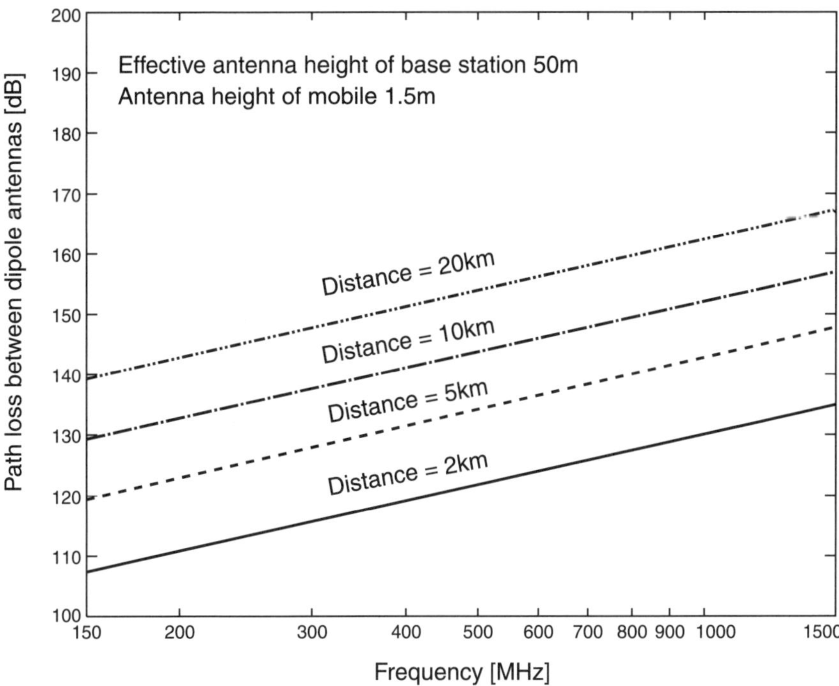

Figure 4.3 Propagation loss estimated by Okumura-Hata method.

The propagation loss estimated by the Okumura-Hata method is shown in Figure 4.3 as a function of frequency. A coverage area with a radius of more than 1 km of a base station can be obtained using this propagation loss estimation method.

It is possible to calculate the output power and antenna gain of the base station from the propagation loss estimated by the Okumura-Hata method. If P_t, G_t, and M_s represent the transition power, antenna saturation gain, and system margin, respectively, the received electric power P_{er} at the cell edge can be formulated as follows:

$$P_{er} = E_{er} + P_t + G_t - L_p - M_s \quad \text{dB} \tag{4.6}$$

The system margin is between 10 and 20 dB and is caused by the unexpected leveling of the received electric field strength due to fading. The noise level at the location of the base station determines the cell edge of a cellular system. The noise level in the frequency range of 800 to 2000 MHz is between

0 and 10 dBμV and can be converted into the unit of dBm by the conversion equation shown in Appendix A, and is as follows:

$$P_r(\text{dBm}) = V_e(\text{dB}\mu\text{V}) - 113.01 \tag{4.7}$$

where a feed line having an impedance of 50Ω is assumed.

The propagation loss at 900 MHz can be computed where the height of the base station antenna h_b = 30m and the height of the mobile station h_b= 1.5m. If the cell radius of the base station in big cities is assumed to be 3 km, the propagation loss L_p becomes 136 dB from (4.1) through (4.3). If the minimum electric field strength received at the cell edge is 10 dBμV = 103 dBm and the system margin M_s = 20 dB, then the following equation applies:

$$P_t + G_t = E_{er} + L_p + M_s = -103 + 136 + 20 = 53 \quad \text{dB} \tag{4.8}$$

An antenna gain of 10 dB is necessary for this estimation if the output power of the base station is 20W, which is equivalent to 43 dBm. Based on this procedure and other factors such as the coverage area of the base station and the gain of the base station antenna, the transmitting output can then be determined.

4.1.2 Breakpoint of a Microcell

Okumura's curve was derived for early land mobile communication systems for which the coverage area of the base stations was assumed to be from several kilometers to 10 km in radius. Therefore, it is difficult to apply to a case where the coverage area is below 1 km from a base station, such as occurs in a microcellular system, where the purpose is to increase the capacity and high-speed data transmission capability. Generally, the propagation loss of microcells is based on line-of-sight (LOS) propagation (defined in Section 1.2.1), and assumes a breakpoint where the propagation loss changes rapidly [4, 5].

The breakpoint can be explained by the two-wave model, by considering the direct path and the ground reflection path. The Earth is approximated as a flat surface, as shown in Figure 4.4, where the propagation distance is denoted by d_1 for the direct path between transmission and reception point, and d_2 for the ground reflection path.

When the heights of the transmitting point h_b and the receiving point h_m are very small compared with the distance d between the transmitting and receiving points, the distance in these two paths is given by the following:

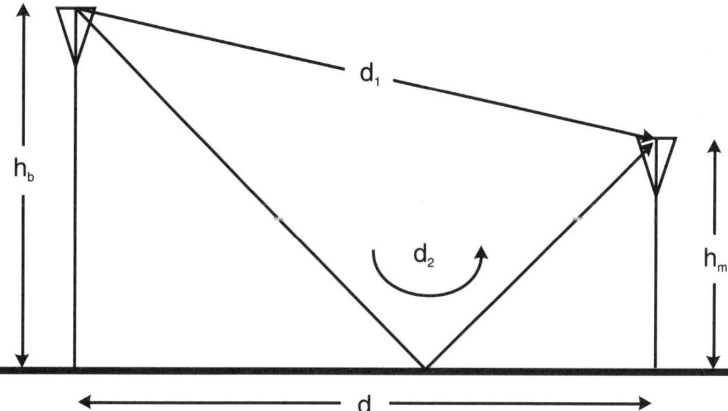

Figure 4.4 Two-wave model with the flat Earth.

$$
\begin{aligned}
d_2 - d_1 &= \sqrt{d^2 + (h_b + h_m)^2} - \sqrt{d^2 + (h_b - h_m)^2} \\
&= d\left\{1 + \frac{1}{2}\left(\frac{h_b + h_m}{d}\right)^2\right\} - d\left\{1 + \frac{1}{2}\left(\frac{h_b - h_m}{d}\right)^2\right\} \cong \frac{2h_b h_m}{d}
\end{aligned}
\tag{4.9}
$$

The propagation loss increases rapidly when the distance given by (4.9) is more than a half-wavelength, because a reflection wave from the ground then cancels a direct wave, due to the phase difference becoming equal to π. The point where the propagation distance satisfies the above condition is called a *breakpoint,* and the corresponding distance, $d_2 - d_1$, is defined as:

$$
b = \frac{4h_b h_m}{\lambda} \tag{4.10}
$$

However, because fading causes large local fluctuations in the measured electric field, the position of the breakpoint cannot be clearly determined from measurements of the propagation loss. Assume that a breakpoint exists in an area between 100 and 500m from a base station, and that an example measurement result is as shown in Figure 4.5. From measurements at frequencies of 457.3 MHz, 2.2 GHz, and 4.7 GHz along a checker-shaped road in the city, the breakpoint position b_m is given as [6]:

$$
b = \frac{1.9h_b h_m}{\lambda} \tag{4.11}
$$

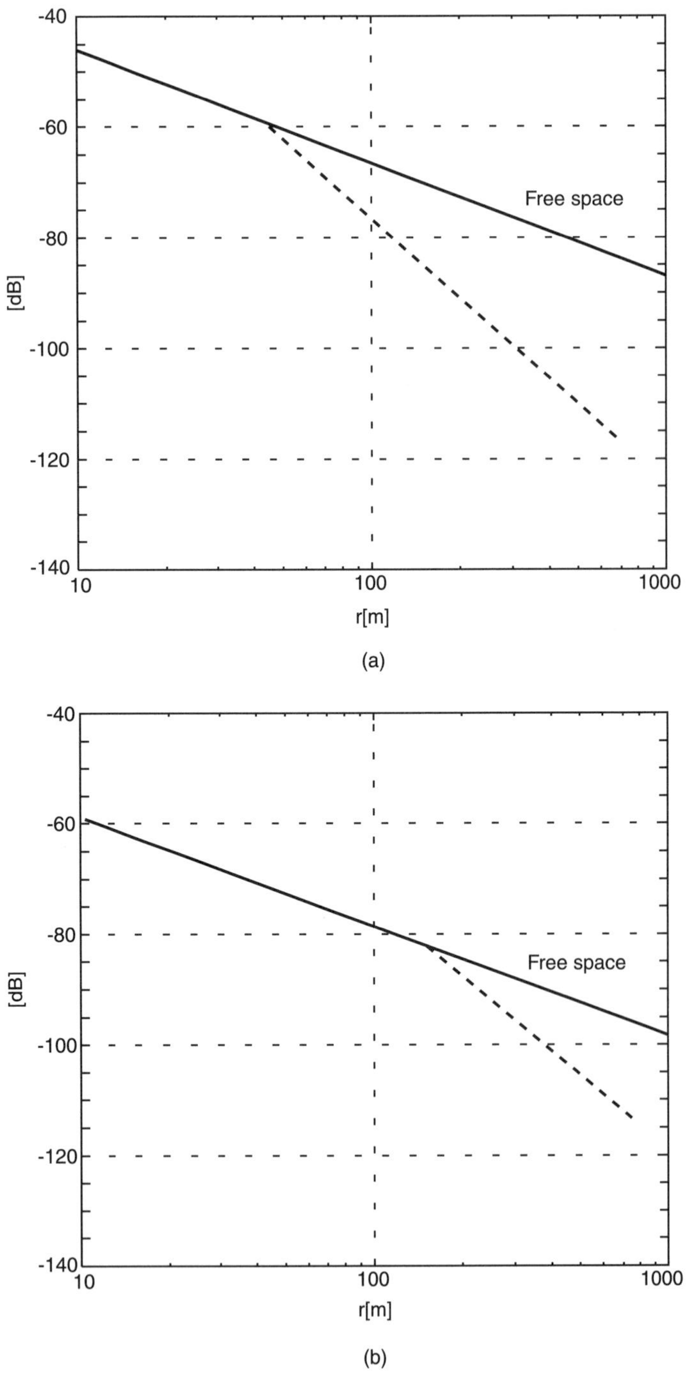

Figure 4.5 Propagation loss and breakpoint: (a) 457.3 MHz, (b) 2.2 GHz, and (c) 4.7 GHz.

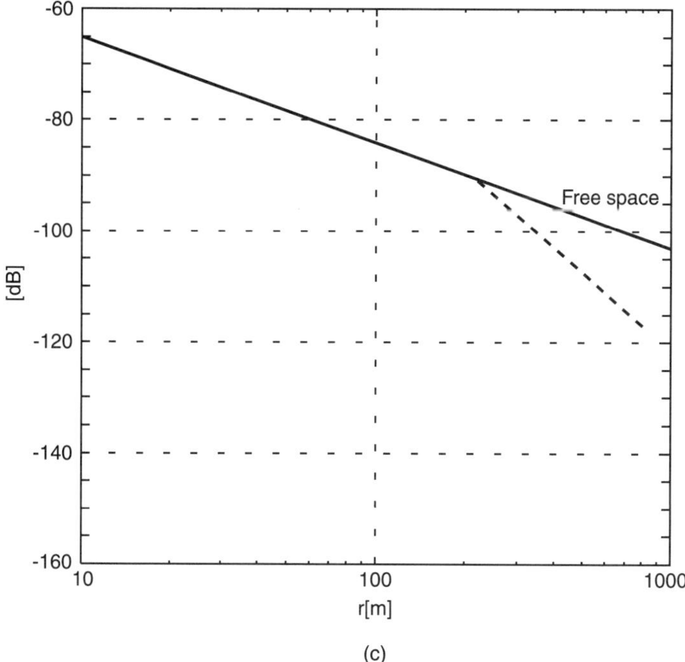

Figure 4.5 Continued.

Quoted propagation loss estimations are based on data from measurements at 457.3 MHz, 2.2 GHz, and 4.7 GHz in Yokohama, where main roads are checker shaped. This reference has been selected because the frequency range is broad and the approximated closed-form equation expresses the propagation loss before and after the breakpoint. The propagation loss in the LOS region is inversely proportional to the distance *r*. It depends on the propagation loss in free space before the breakpoint and attenuates according to $1/r^4$ after the breakpoint. If the same expression as the Okumura curve is taken as the propagation loss, it can be formulated by denoting frequency as f MHz and K as a constant before the breakpoint ($d < b$) as:

$$L_{bb} = -20 \log d - 20 \log f + K \quad \text{dB} \tag{4.12}$$

and after the breakpoint, as:

$$L_{ba} = -40 \log\frac{d}{b} + L_{bb} = -40 \log d + 20 \log(1.9h_b h_m) - 20 \log c + K \text{ dB} \tag{4.13}$$

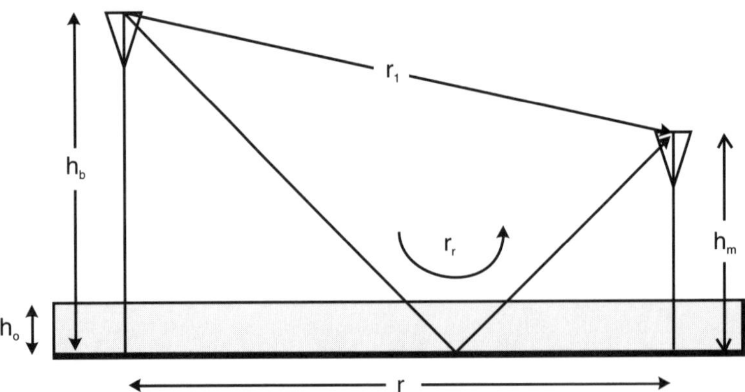

Figure 4.6 Two-wave model using effective Earth surface.

where c is the velocity of light and the frequency term is excluded by assuming variation as $1/r^4$ for $d > b$. The value of K is 27.6 in this example.

The value of propagation loss estimated by the two-way model is smaller than that obtained from the measured results because of the effective raising of the Earth surface by vehicles on the road and the blocking of the propagation path by signboards and pedestrian overpasses. These two factors are shown in Figure 4.6 and are introduced to the propagation loss as:

$$L_d(R) = e^{-sr} \frac{\lambda}{4\pi} \left| \frac{e^{-jk_o r_1}}{r_1} + R(\alpha) \frac{e^{-jk_o r_m}}{r_{rm}} \right| \tag{4.14a}$$

$$r_{rm} = \sqrt{r^2 + \left\{(h_b - h_o) + (h_m - h_o\right\}^2} \tag{4.14b}$$

In the results from measurements carried out in Sapporo, Japan, the preceding estimate is reported to agree well when values of $s = 0.001$, and $h_o = 1$m are used for daytime and $h_o = 0.5$m for nighttime [7]. This formula can estimate the propagation loss without dividing the path by the breakpoint.

4.1.3 Propagation Loss Outside the Line of Sight

In a propagation model for a microcellular system, there can be a road in a direction that curves away from a corner and constitutes a region that is outside the LOS. Beyond the corner, the propagation paths are multiple reflected and diffracted waves, as shown in Figure 4.7. The effects of the diffracted

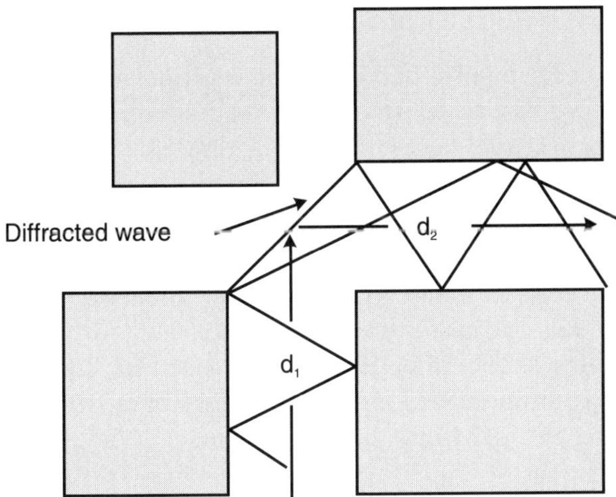

Figure 4.7 Propagation path at intersection.

waves are small and can be neglected. The propagation loss can be divided into the loss at the corner and that beyond the corner. In general, the loss at the corner varies from 10 to 30 dB depending on the propagation conditions and on the distances between the transmitting and receiving points and the corner. The loss can be expressed in terms of the distance d_2 from the corner to the receiving point as:

$$L_{nl} = \alpha \log d_2 + K_n \quad \text{dB} \tag{4.15}$$

Reference [6] indicates that α has little frequency dispersion and its recurrence curve can be approximated by the measurement data as:

$$\alpha = 14.5 \log d_1 - 51.0 \quad \text{dB} \tag{4.16}$$

where d_1 is the distance from the transmitting point to the corner.

A further method for estimating the propagation loss uses the total distance $r = d_1 + d_2$ without considering the differences for the regions inside or outside the LOS. In this case, the propagation loss is the same as in free space for distances up to 30m away from the transmitting point, and is proportional to $1/r^5$ from there on. The corresponding expression does not then depend on the location of the breakpoint, and is as follows:

$$L_{nl} = -50 \log d - 20 \log f + 30 \log 30 + K \quad \text{dB} \tag{4.17}$$

4.1.4 Propagation Loss Estimation by the Ray-Tracing Method

The results of a large number of propagation experiments have been required to derive the approximate equation for the estimation of propagation loss. Recently, the propagation loss, particularly within microcellular systems, has been estimated by numerical simulation using the ray-tracing method. The ray-tracing method approximates electromagnetic waves by optical rays, and is applicable for simulation of reflection, diffraction, and scattering by buildings and other obstacles. However, it is difficult to model all the real influences precisely. For example, it is impossible to include all the influences of trees on a street, windows in buildings, window frames, and active persons inside buildings in the calculation model. However, accurate modeling is indispensable when considering the precise changes in the electric field level at the receiving point.

On the other hand, a relatively simple calculation model can give the median level of electric field strength at the receiving point and the average propagation loss [8]. For example, when a building can be considered high in comparison with the height of the base station antenna, the propagation loss along a street in a large city can be approximated to that in a U-shaped guide, for which all the surfaces are concrete walls, as shown in Figure 4.8. The electric field strength at the receiving point in the U-shaped guide is calculated as the sum of the direct waves from the transmitting point and the reflected waves from the walls. A calculated result obtained using such an approximate model is shown in Figure 4.9. The height of the base station h_m, the radiation pattern of the base station antenna, and other charac-

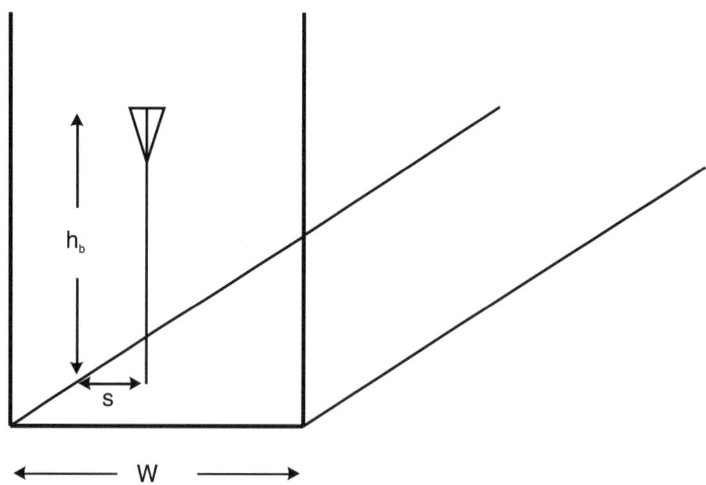

Figure 4.8 U-shaped guide for ray-tracing simulation.

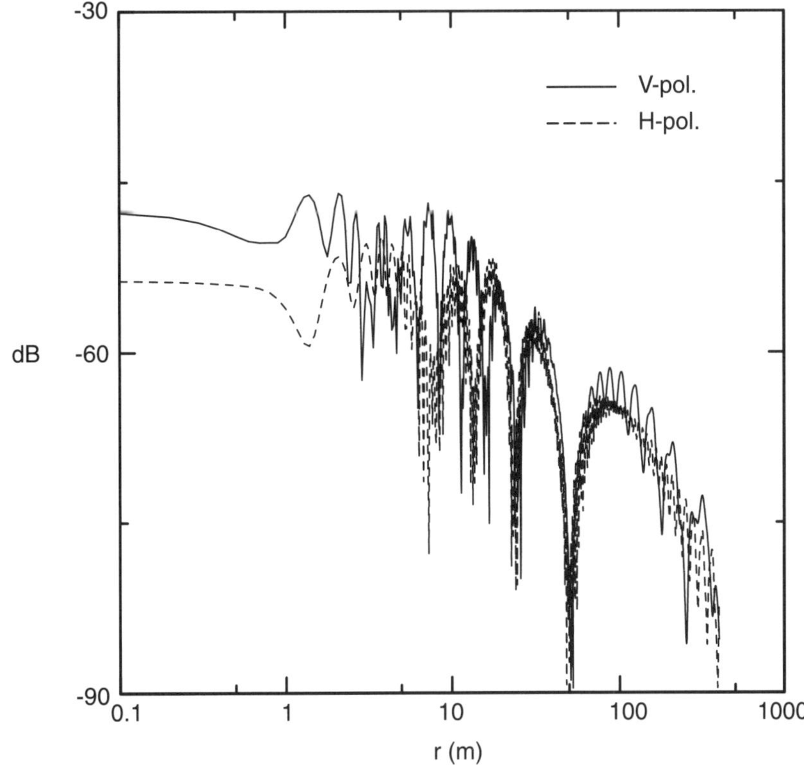

Figure 4.9 Calculated electric field distribution from ray-tracing method: $f = 1.9$ GHz, $W = 10$, $S = 0.1$, $h_b = 2.5$, $h_m = 1.5$(m).

teristics can be calculated, and it is therefore an effective technique for estimating the propagation phenomena of microcellular systems.

4.2 Design of Cellular Base Station Antennas

This section presents the base station antenna structure including the feeding network, and a synthesis method for pattern shaping to give the desired characteristics. The uplink diversity scheme is mandatory for base station systems, and the design procedures for space and polarization diversity are also described in this section.

4.2.1 Antenna Array Structure and Feeding Circuit

The horizontal radiation pattern is often omnidirectional; however, a sector-shaped pattern is sometimes used to increase the capacity of subscribers. This sector system covers the whole of the horizontal plane by dividing it into several sectors. The sector pattern is given by the antenna elements supplemented by reflector or parasitic elements in front of the antenna, as shown in Figure 4.10. A narrow sector pattern with a 60° half-power beamwidth requires a two-element array in the horizontal direction, excited in phase.

The antenna radiation pattern in the vertical plane consists of a shaped beam to control the coverage area of the base station and constitutes an array antenna in the vertical direction. A simple pattern used for cellular systems is a pencil beam to increase the directivity gain in the vertical direction. This pattern is given by an array in which the phases and amplitudes of all the elements are the same.

A tournament-feeding network, shown in Figure 4.11(a), provides the in-phase excitation. This feed network is useful for high-gain antennas of more than 15 dBi, although an antenna gain of about 10 dBi can be obtained with the colinear array shown in Figure 4.11(b). This colinear array consists of several half-wavelength coaxial cables connecting the inner conductor of the lower element to the upper outer conductor in order to change the phase alternatively at the edge. Although this antenna structure is very simple, the input impedance is not matched at the frequency of the optimum structure

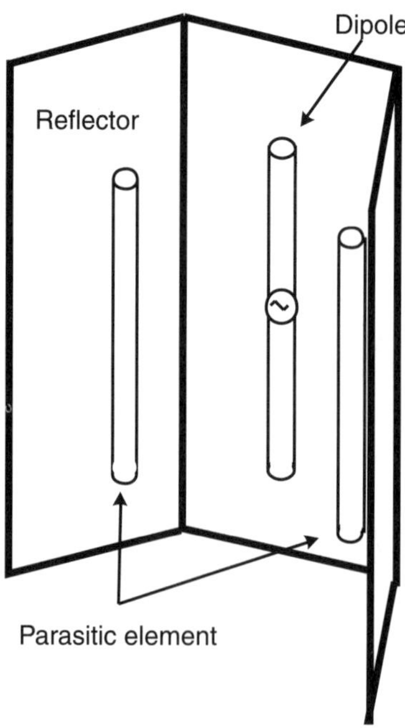

Figure 4.10 Sector beam antenna using a reflector and a parasitic element.

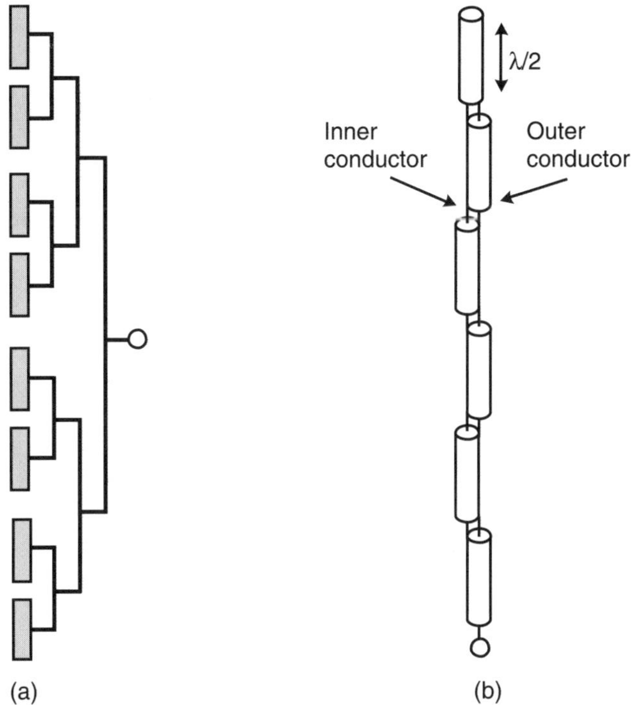

Figure 4.11 Feeding network for base station antenna array: (a) tournament array and (b) colinear array.

parameter for the required radiation pattern. Impedance matching is then obtained by matching the circuit at the feed point.

Typical antenna elements for cellular systems are printed dipole, microstrip, and printed slot, as shown in Figure 4.12. The antenna element is selected according to the design of the base station and its cost. The feeding network is usually made from microstrip line.

4.2.2 Beam Tilting and Propagation Characteristics

To increase the subscriber capacity in current cellular systems, a smaller coverage area is required for each base station. In the frequency-division multiple-access (FDMA) system, the same frequency can be used in several cells. A restriction in coverage area is also effective for the code-division multiple-access (CDMA) system, because the increase in the number of users in the same cell raises the base noise level. Therefore, a similar cell form is required for the CDMA system as for the FDMA system. To restrict the cell

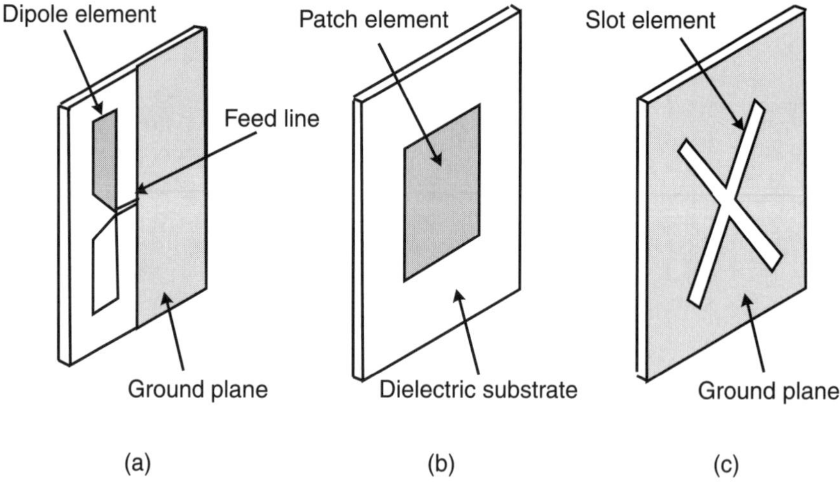

Figure 4.12 Antenna elements for base station antennas: (a) printed dipole, (b) patch, and (c) slot.

size, the beam tilting technique shown in Figure 4.13 is effective. In the FDMA system, the null position of the antenna pattern is directed toward another cell in order to minimize the interference between cells using the same frequency band.

Using tilted beams, the propagation loss becomes larger than without tilting, as shown in Figure 4.14, which also shows the effectiveness of using tilted beam antennas.

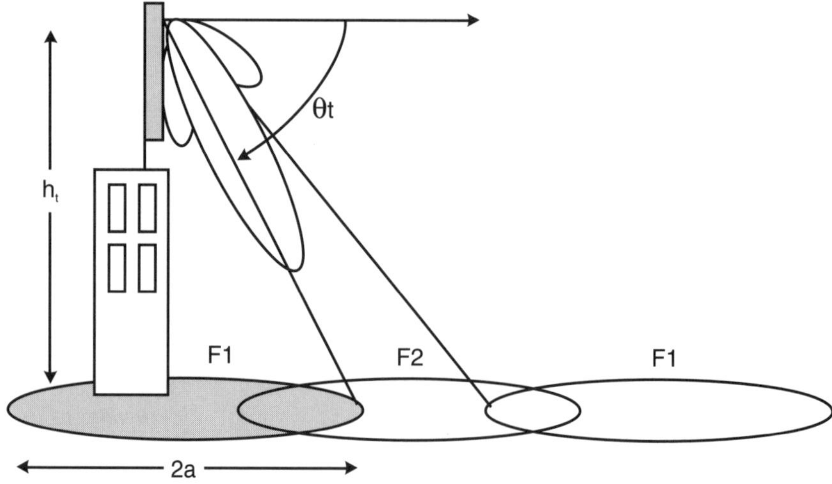

Figure 4.13 Beam tilting effect of base station antenna.

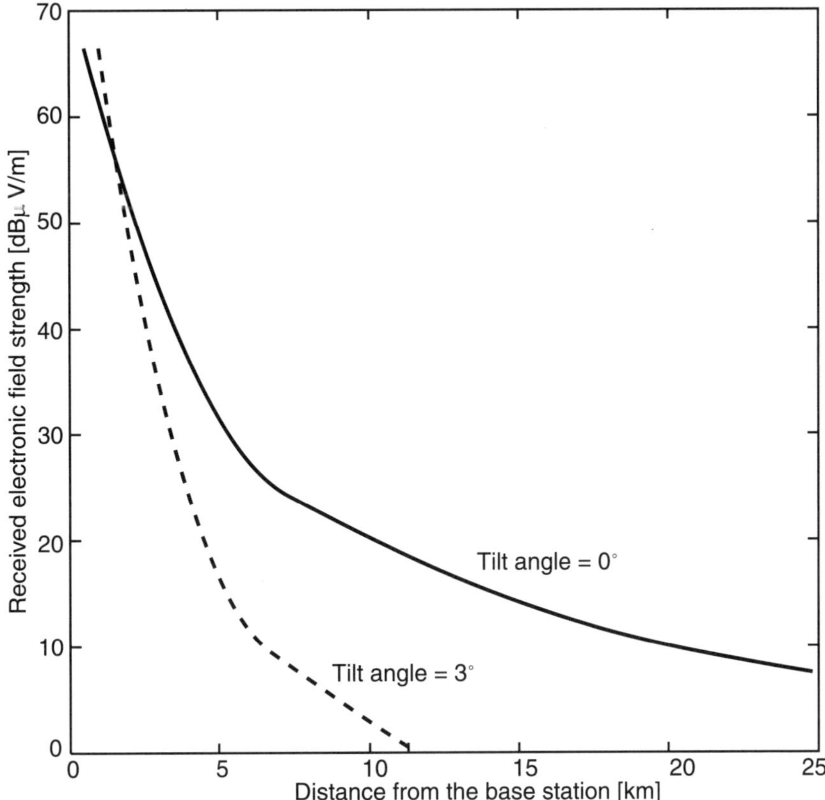

Figure 4.14 Propagation loss by antenna main beam tilting.

For a system in which the antenna height of the base station is h (in meters) and the cell radius is a (in meters), the tilt angle θ_t can be determined as:

$$\theta_t = \tan^{-1}(h/a) \tag{4.18}$$

For example, the tilt angle is $\theta_t = 3.8°$ for a system in which $h = 100$m and $a = 1.5$ km. In a practical design, which includes the interference margin between adjacent cells, the real tilt angle is 1 to 2° larger than the value given by (4.18).

A design example of the vertical plane pattern for a base station antenna is shown in Figure 4.15. Denoting $\theta = 0°$ as the horizon, the lower region $0 < \theta < 90°$ is a pattern corresponding to $\text{cosec}^2\theta$, and the upper region $90° < \theta < 0°$ is designed as a Taylor distribution specifying the sidelobe level [9]. In particular, the level in the direction of another cell using the same fre-

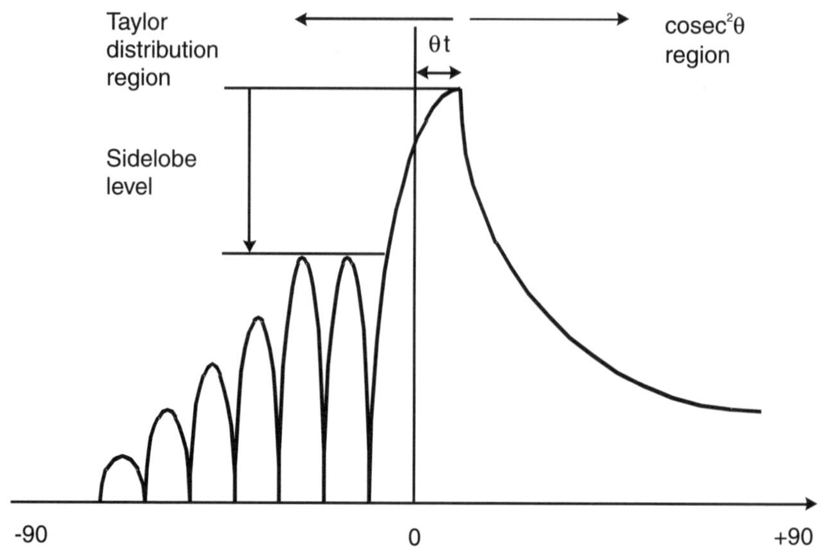

Figure 4.15 Example of vertical plane pattern for base station antenna.

quency is often specified as very low. This design is based on antenna pattern synthesis, and the excitation condition of each antenna element is evaluated by fixing the element spacing as described below.

For a design example, this section presents the least squares method for pattern synthesis. If the desired radiation pattern is $g_d(\theta)$, as shown in Figure 4.16, the pattern given by the actual feeding network of $g(\theta)$ can be expressed by the amplitude I_i and phase ϕ_i as:

$$g(\theta) = f(\theta)\sum_{n=0}^{N-1} I_i \exp(jnk_0 d \cos\theta + j\phi_n) \quad (4.19)$$

where N and d represent the number of arrays and the element spacing, respectively, and k_0 is the wave number in free space.

To obtain the amplitude I_i and phase ϕ_i for the desired antenna pattern, the error value $\delta = |g_d(\theta) - g(\theta)|$ should be minimized. Discreet points of M, in the following vector, express the desired pattern in the vertical plane:

$$G_d = \begin{bmatrix} g_{d1}(\theta_1) \\ \vdots \\ g_{dM}(\theta_M) \end{bmatrix} \quad (4.20)$$

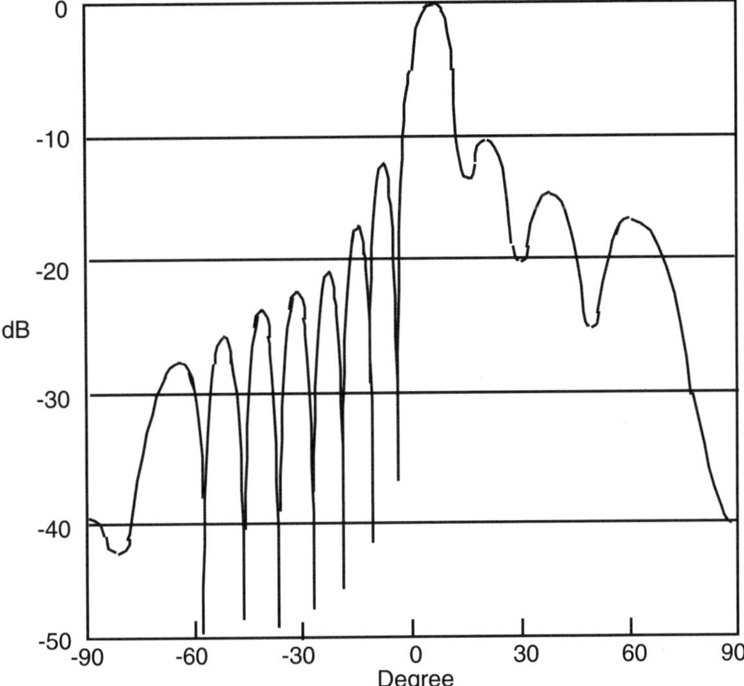

Figure 4.16 Desired radiation pattern in vertical plane.

Using (4.19), $g(\theta)$ can be given in matrix form from $A_{mn} = \exp(jnk_0 d\ \sin\theta_m)$ and $a_n = I_n e^{j\phi_n}$ as:

$$G = \begin{bmatrix} g_1 \\ \vdots \\ g_M \end{bmatrix} = \begin{bmatrix} A_{11} & \cdots & A_{1N} \\ \vdots & \ddots & \vdots \\ A_{M1} & \cdots & A_{MN} \end{bmatrix} \begin{bmatrix} a_1 \\ \vdots \\ a_N \end{bmatrix} \tag{4.21}$$

If the error function to minimize the value of δ is ε, which is defined as:

$$\varepsilon = \sum_{m=1}^{M} \left(g_{dm} - \sum_{n=1}^{N} A_{mn} a_n \right) \left(g_{dm} - \sum_{n=1}^{N} A_{mn} a_n \right)^* \tag{4.22}$$

The condition for the excitation coefficient of a_n to minimize ε is then given by the least squares method as:

$$\frac{\partial \varepsilon}{\partial a_n^*} = \sum_{n=1}^{N} \left(g_{dm} - \sum_{n=1}^{N} A_{mn} a_n \right) A_{mn}^* = 0 \tag{4.23}$$

4.2.3 Sector Antennas

The sector zone allocation requires a fan-shaped pattern in the horizontal plane. However, an ideal fan shape cannot be obtained from a real antenna and the sector region is specified by the half-power beamwidth. This real sector pattern can be approximated by denoting the half-power beamwidth as ϕ_b [10], where:

$$g(\theta) = 10m \log_{10}\left(\cos(\phi/2)\right) \quad -\phi_e \le \phi \le \phi_e \quad \text{dB} \tag{4.24a}$$

$$= -\text{FB} \quad -\pi \le \phi \le \phi_e,\ \phi_e \le \phi \le \pi \quad \text{dB} \tag{4.24b}$$

$$m = \frac{-3}{10 \log_{10}\left(\cos(\phi_b/2)\right)}, \quad \phi_e = 2 \cos^{-1}\left(10^{-\frac{2}{m}}\right) \tag{4.24c}$$

Current cellular systems use three-sector zones, where the half-power beamwidth is 90° or 120°. The beamwidth depends on the system design. Although a narrow beam increases the user number, it sometimes causes blind regions at the cell edge. Therefore, the base station location is important. For the next generation of mobile communication systems—International Mobile Telecommunications (IMT-2000)—six sectors will be used because such an arrangement gives double the capacity of the three-sector system [11].

4.2.4 Diversity Antennas

Uplink diversity reception is mandatory for base station antennas in order to decrease multipath fading. The diversity reception described in Section 1.3.1 is also used to compensate for power imbalances between uplinks and downlinks, because the uplink signal is weaker than that of the downlink. For a space diversity system, the same antenna is installed with a spacing of more than 10 wavelengths (λ). The correlation coefficient for this space diversity, as shown in Figure 4.17, is calculated by assuming that the uplink

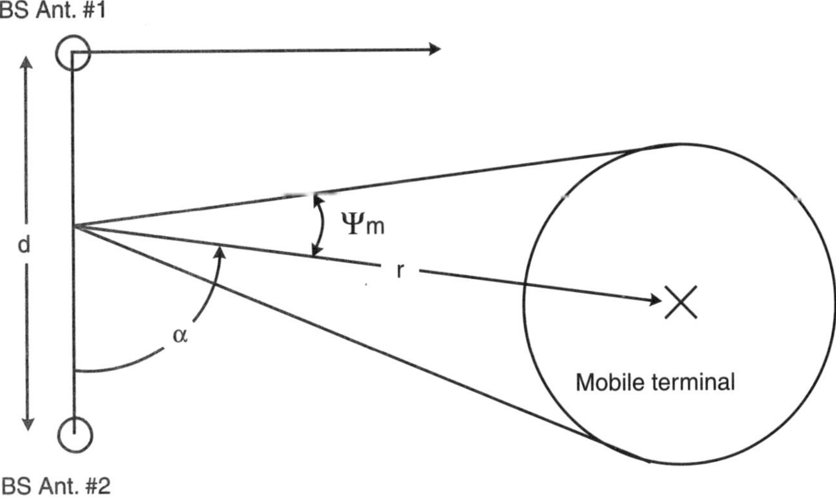

Figure 4.17 Coordinate system for space diversity antenna.

signals originate from inside the angle ψ_m with uniform distribution. The correlation coefficient of ρ_e is evaluated from the definition of (2.32) as:

$$\rho_e = \left(\frac{\sin X}{X}\right)^2, \quad X = 2\pi\left(\frac{\Psi_m d}{\lambda}\right)\sin\alpha \tag{4.25}$$

The calculated correlation is shown in Figure 4.18. For a design value of $\rho_e < 0.5$ for diversity reception, the antenna spacing of d should be more than 10λ, because $\psi_m > 1.5°$ is obtained from propagation measurements in the cities [12].

Mobile telephone services started as vehicular-mounted mobile terminals. Recent developments in electronics technology have resulted in very small and lightweight portable handsets. As a result, everyone now uses a handset that is held near the ear at a tilted angle. As derived from statistical data [13], the tilt angle is about 60°, and this decreases the vertical radiation component by 6 dB, and the horizontal component by 5 dB more than the vertical component. This operating condition results in an increase in the horizontal electric field component in the uplink, and the polarization diversity is effective for the diversity system.

Vaughan measured a polarization diversity scheme in the 400-MHz band using a tilted dipole as a mobile terminal [14] and showed the correla-

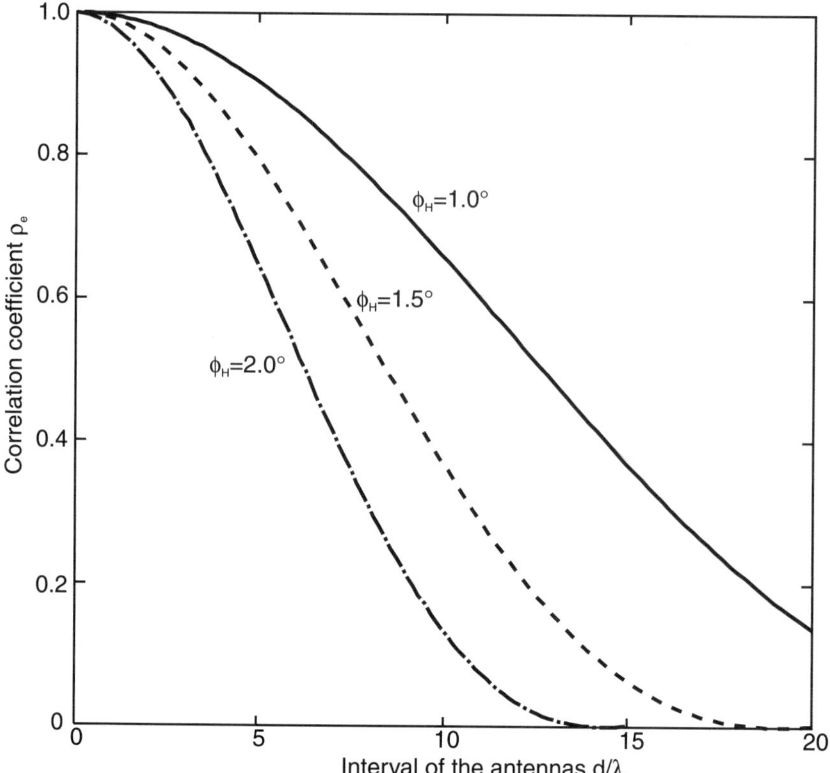

Figure 4.18 Correlation coefficient of space diversity antenna.

tion coefficient between horizontal and vertical components to be almost zero in both the city and the suburbs. Propagation experiments by Turkmani at 1.8 GHz resulted in a diversity gain of 1 to 2 dB for vertically oriented dipole antennas as the mobile terminal and 3 to 5.2 dB for 45° tilted antennas [15]. The measured correlation coefficient quoted by Lotse at 1.8 GHz was less than 0.2 [16], and the diversity gain depended on the average received signal level. The polarization diversity gain measurement for the case with a human operator using a handset was 7 dB greater than the space diversity in the LOS region, and 1 dB greater outside the LOS region [17]. The polarization diversity system using vertical and horizontal components has the merit of minimizing the antenna installation space, and has been adopted for Japanese cellular systems.

Another polarization diversity system uses slant polarization of ±45°. The correlation coefficient is about 0.5, but the lack of unequal received signal levels for both diversity ports gives a large diversity gain that is the same as that for vertical and horizontal diversity systems.

4.3 Base Station Antenna Measurements

Measurements on the base station antenna are classified into low-power tests and high-power tests. In the lower power tests, the antenna radiation pattern and the input impedance characteristics are measured. A procedure similar to that used for the measurement of the antenna can also be used to measure the input impedance of the handset terminal, as presented in Chapter 2. It is necessary for the pattern measurement of a base station antenna to be an open site or near-field measurement when the antenna has a high gain, so it cannot be measured in an anechoic chamber. Because beamwidth and the position of the null point should be estimated precisely for the designed coverage area of the cell, this section presents a measurement method for high-gain antennas that cannot be measured in an anechoic chamber. In any case, the passive intermodulation (PIM) causes serious problems, especially when the base station antenna is operated at high power levels. This section describes the definition of PIM and its measurement. Finally, the section also presents the inspection process of the base station antenna as a product.

4.3.1 Antenna Pattern Measurements in Open Sites

The distance R between the transmitting and receiving antennas should satisfy the condition $R \geq 2D^2/\lambda$, defined by (2.11), to measure the far-field radiation, as presented in Chapter 2, where the maximum dimension of the antenna is denoted as D. For a cell radius of less than 3 km, the base station antenna controls the phase characteristics of the array elements electrically to vary the radiation pattern in the vertical plane. Because the dimension in the vertical direction becomes more than 10λ for the pipe-shaped form, and the measurement distance R requires more than 200λ as a condition of far field, in the 900-MHz band, a distance of more than 67m is necessary, which is difficult to obtain in an indoor facility. Measurements must therefore be conducted in an open site, and the transmitting antenna must be installed on a steel tower or a building, as shown in Figure 4.19, in order to satisfy the above far-field condition.

For radiation pattern measurements, it is necessary to prevent interference from the radio link that is actually used and to suppress reflections from surrounding buildings. To avoid receiving waves reflected from the ground,

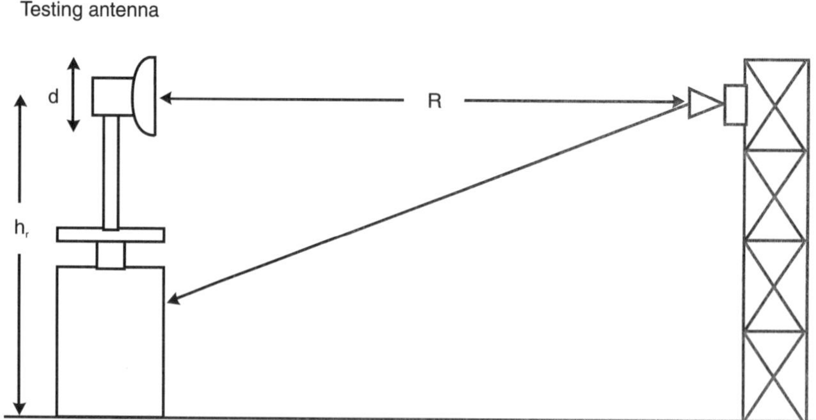

Figure 4.19 Far-field measurement using a tall tower.

the antenna aperture size d should satisfy the following equation, where h_r is the height of the receiving antenna (the antenna used for the measurements) and R is the distance between the transmitting and the receiving antennas:

$$d \geq \frac{1}{2} \times 0.886 \frac{\lambda R}{2h_r} \tag{4.26}$$

The antenna aperture size d is defined as the largest dimension of the test antenna. If the aperture size of the transmitting antenna is much greater than the wavelength, and the antenna aperture is assumed to have a uniform current distribution, the half-power beamwidth $2\theta_h$ of the transmitting antenna can be approximated by the following equation. This equation was derived so that the edge of the half-power beam points to the tower base of the receiving antenna:

$$2\theta_h \cong 0.886 \frac{\lambda}{d} \tag{4.27}$$

When the required distance is not available for the measurements, a diffraction fence can be installed on the central ground between the transmitting and the receiving antennas to avoid the influence of the waves reflected from the ground. The diffraction fence should be mounted so as not to interrupt the main beam between the transmitting and receiving antennas, and it should be a saw-shaped fence in order to give knife edge diffraction.

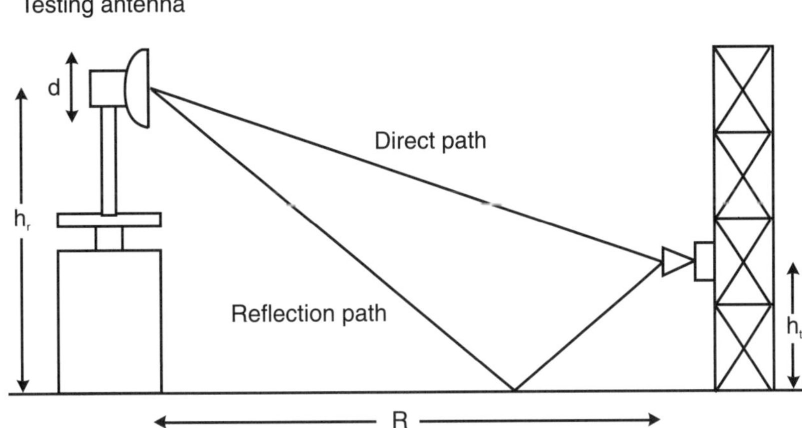

Figure 4.20 Far-field measurement using ground reflection.

A measurement method that uses the reflection from the ground actively is shown in Figure 4.20. The antenna under test is installed near the ground, and its height is adjusted so that the maximum electric field strength is received. In this method, it is assumed that the reflection coefficient at the ground has a value close to –1 when both the transmitting and the receiving antennas are located just above the ground. Measurements can be made in similar conditions by adjusting the height for an arbitrary frequency. Based on the condition that the phase difference in the image antenna becomes $(\lambda/2)$, the height h_r of the antenna under test can be determined as:

$$h_r = \frac{\lambda R}{4h_t} \tag{4.28}$$

For the slant range measurement, the antenna under test is installed on the top of a high steel tower, as shown in Figure 4.21, and the measurement antenna on the ground is pointed toward it with a large elevation angle. The reflection waves from the ground are radiated from the direction of the sidelobes, and this is an effective method for determining the direct path component in the measurement. However, the waves reflected from the surrounding buildings need to be avoided.

4.3.2 Near-Field Antenna Pattern Measurement

The problems associated with open site measurements include noise from the circumference of the site, the influence of weather, and the requirement for

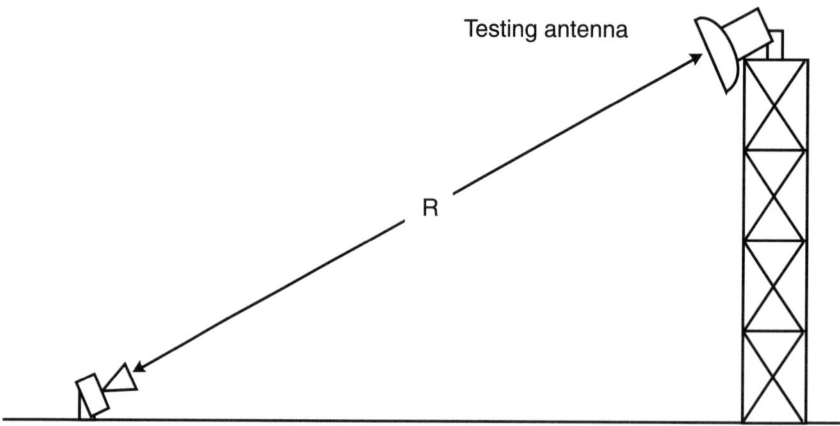

Figure 4.21 Far-field measurement using slant range.

a license to transmit electromagnetic waves at high power levels. A technique for making measurements on high-gain antennas in a stable condition is the near-field measurement described in this section.

4.3.2.1 Vector Kirchhoff Integral and Wave Spectrum Expansion

If a virtual boundary S encloses a test antenna, the equivalent electric and magnetic currents **J** and **M** are given by the tangential magnetic and electric fields on the surface S according to the expressions:

$$\mathbf{J} = \mathbf{n} \times \mathbf{H}_t \qquad \mathbf{M} = \mathbf{E}_t \times \mathbf{n} \tag{4.29}$$

The vector Kirchhoff integral given below gives the antenna far field, using the measured tangential magnetic and electric field, as:

$$\mathbf{E}(\mathbf{r}) = \frac{jke^{-jkr}}{4\pi r}\mathbf{r} \times \int_S (\mathbf{M} + Z_o\mathbf{r} \times \mathbf{J})e^{jk\mathbf{r}\cdot\mathbf{r}'}\, dS' \tag{4.30}$$

where Z_o is the characteristic impedance in free space. The procedure requires calibrated, ideal probes for the measurement of both the tangential electric and magnetic fields over the surface S. However, the computation time required to perform the numerical integral in (4.30) is very large [18].

A different approach that reduces the computation time is given by wave vector expansion and the fast Fourier transform (FFT). The electric field radiated by the antenna is expressed as a summation of plane waves expanded in the wave spectrum domain, as:

$$E(r) = \frac{1}{2\pi} \int_{-\infty}^{+\infty} \int \mathbf{f}(k_x, k_y) \exp(-j\mathbf{k} \cdot \mathbf{r}) \, dk_x \, dk_y \tag{4.31}$$

where **r** and **k** denote the observation point vector and the wave vector, respectively, and are expressed by the unit vectors, $\mathbf{e}_x$, $\mathbf{e}_y$, $\mathbf{e}_z$ in the *x*, *y*, *z* coordinate system as:

$$\mathbf{r} = x\mathbf{e}_x + y\mathbf{e}_y + z\mathbf{e}_z, \quad \mathbf{k} = k_x\mathbf{e}_x + k_y\mathbf{e}_y + k_z\mathbf{e}_z \tag{4.32}$$

The wave number k_z given below, which does not diverge as $z \to \infty$, is:

$$k_z = -j\sqrt{k_x^2 + k_y^2 - k^2} \tag{4.33}$$

where k is the wave number in free space.

If the measurement region in the *x*-*y* plane is denoted by *A*, and the tangential (*x*, *y*) component of the wave vector source and the vector in the region *A* are $\mathbf{k}_T$ and $\mathbf{r}_T$, as shown in Figure 4.22, the following equations can be derived:

$$\mathrm{E}_A(x, y) = \frac{1}{2\pi} \int_A \mathbf{f}_T(k_x, k_y) \exp(-j\mathbf{k}_T \cdot \mathbf{r}_T) \, dk_x \, dk_y \tag{4.34}$$

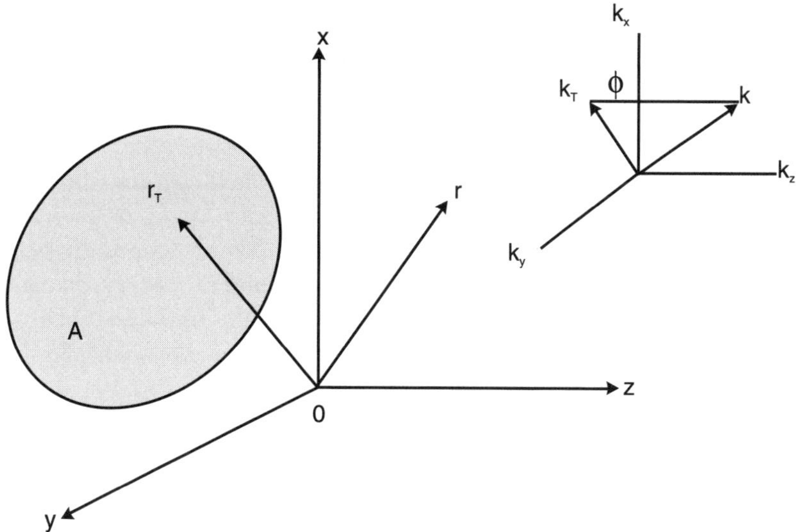

Figure 4.22 Coordinate system for region A and wave vector.

$$\mathbf{f}_\mathrm{T}(k_x, k_y) = \frac{1}{2\pi}\int_A \mathbf{E}_\mathrm{A}(x, y)\exp(j\mathbf{k}_\mathrm{T}\cdot\mathbf{r}_\mathrm{T})\,dx\,dy \qquad (4.35)$$

where the subscript T denotes the tangential (x, y) components. Equation (4.35) indicates that the function of $\mathbf{f}_\mathrm{T}(k_x, k_y)$ can be evaluated by measuring the tangential electric field components in region A.

After evaluation of $\mathbf{f}_\mathrm{T}(k_x, k_y)$, the antenna far field can be calculated using (4.31), assuming the observation vector $\mathbf{r}$ to be sufficiently large. The variable transformation below rewrites (4.31) as:

$$\mathbf{E}(\mathbf{r}) = \frac{1}{2\pi}\int_0^{\pi/2}\int_0^{2\pi} \sin\alpha\cos\alpha\,\mathbf{f}(k_x, k_y)\frac{\exp(-jkr)}{r}\,d\alpha\,d\beta \qquad (4.36)$$

$$x = r\sin\theta\cos\phi, \quad y = r\sin\theta\sin\phi, \quad z = r\cos\theta \qquad (4.37)$$

$$k_x = k\sin\alpha\cos\beta, \quad k_y = k\sin\alpha\sin\beta, \quad k_z = k\cos\alpha \qquad (4.38)$$

where the outgoing wave, $k_z \geq 0$, is only considered in the variable transformation. When kr is large enough for the antenna far-field region, (4.36) can be approximated using the saddle-point method [19], which gives:

$$\mathbf{E}_\mathrm{f}(\mathbf{r}) = jk\cos\theta\,\mathbf{f}_\mathrm{T}(k_x\sin\theta\cos\phi, k_y\sin\theta\sin\phi)\frac{\exp(-jkr)}{r} \qquad (4.39)$$

4.3.2.2 Near-Field Measurement

In antenna near-field measurements, the distance between the antenna and the observation point classifies the electromagnetic fields, as described in the following. In an example of large aperture size, such as can be found in a parabolic reflector antenna, the electromagnetic fields are defined differently, by distance in the vicinity, in the near-field region and in the far-field region, respectively, as shown in Figure 4.23. Accordingly, the equivalent theory for the electromagnetic field in each region gives the radiation field. For field measurements in the vicinity of the antenna, the mutual coupling between the measurement probe and the antenna under test becomes very high and disturbs the electromagnetic field radiated from the antenna. Therefore, measurements in the vicinity of the antenna are very difficult. Although it is possible to exclude this mutual coupling by probe correction techniques,

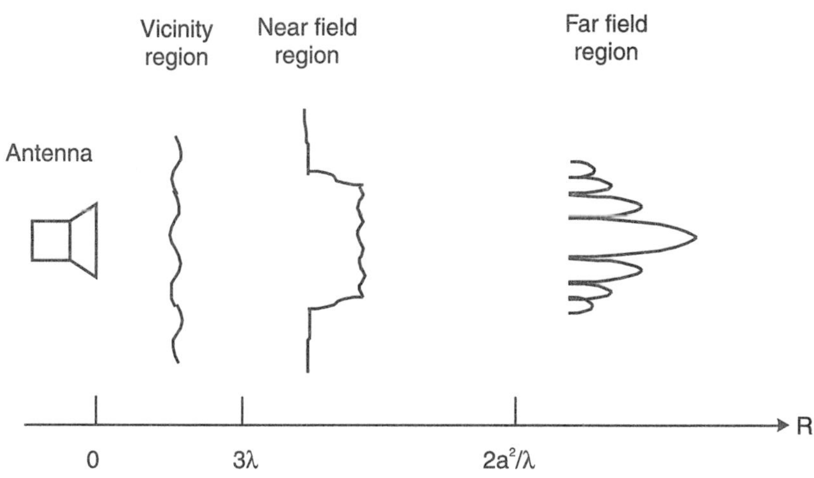

Figure 4.23 Electromagnetic fields in vicinity, near-field, and far-field regions.

measurements in this region are not generally performed because of the instability of the measured field.

Next to the region in the vicinity of the antenna is the near field. A suitable probe for near-field measurement consists of an open-ended rectangular waveguide, which does not have an ideal radiation pattern. This nonideal performance can be rectified easily by probe correction [18], and the far field can then be evaluated from the electromagnetic field in the near field. Although near-field measurements require a space enclosing the antenna, a closed plane is not necessary at the edge of the plane because here the measurements reduce to low values. The probe scanning surfaces for near-field measurements can be plane, cylindrical, or spherical, as shown in Figure 4.24, and correspond to the coordinate system.

Plane surface scanning covers only half the space during the measurements and is appropriate for the measurements on high-gain antennas such as parabolic antennas. Cellular base station antennas and marine radar antennas have fan-shaped patterns with a narrow beam in one plane and a wide beam in the other plane, and cylindrical scanning is appropriate for near-field measurements on these antennas. Spherical scanning has the advantage that low-gain antennas can be measured by scanning enclosed surfaces, but mechanical properties are very severe during scanning.

The scanning area (A × B) and the sampling interval (Δx, Δy) are discussed in the planar scanning method and are shown in Figure 4.25. The size of the scanning area is determined by the requirement to obtain more than a 35-dB dynamic range in the scanning area, without the need to use an

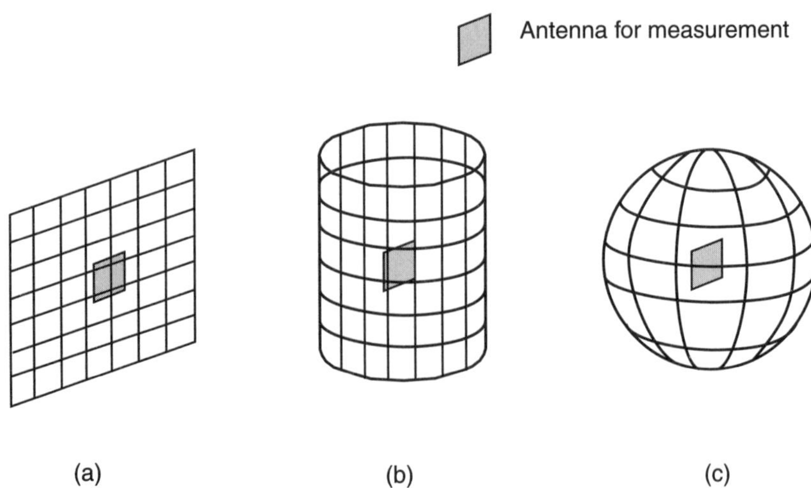

Figure 4.24 Probe scanning surface for near-field measurement: (a) planar, (b) cylindrical, and (c) spherical scanning.

interpolation method on the measured data. In this case, dynamic range is defined by the difference between the maximum electric field strength and the edge field strength. The maximum data point spacing should be such that $\Delta x = \lambda_o/2$ and $\Delta y = \lambda_o/2$ for planar scanning, using the sampling theorem [18].

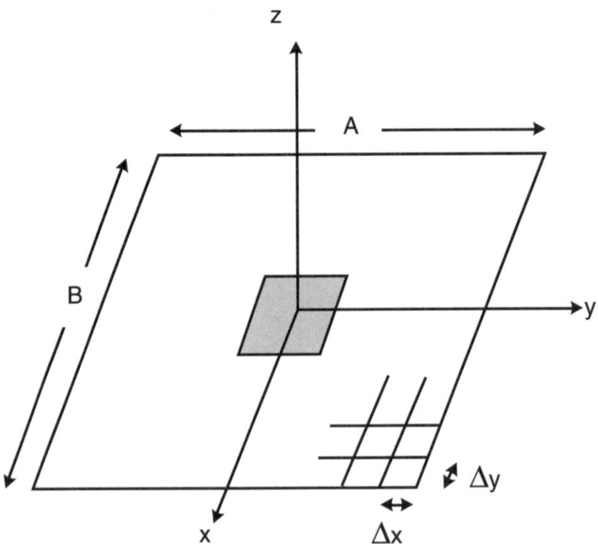

Figure 4.25 Scanning area size and sampling interval for planar scanning.

4.3.2.3 Far-Field Evaluation in Near-Field Measurements

If the x and y components of the measured electric field are $E_x(x, y)$ and $E_y(x, y)$, respectively, the far-field radiation patterns are given by (4.34) and (4.35) and the coordinate conversion from (x, y, z) to (r, θ, ϕ) as:

$$E_\theta = jk \cos\theta \frac{\exp(-jkr)}{r}(S_x \cos\theta \cos\phi + S_y \cos\theta \sin\phi) \qquad (4.40)$$

$$E_\phi = -jk \cos\theta \frac{\exp(-jkr)}{r}(S_x \sin\phi - S_y \cos\phi) \qquad (4.41)$$

$$S_i = \frac{1}{2\pi}\int_{-B/2}^{B/2}\int_{-A/2}^{A/2} E_{Ai}(x_o, y_o)$$
$$\exp\left\{jk_o(x_o \sin\theta \cos\phi + y_o \sin\theta \sin\phi)\right\} dx_o\, dy_o, \quad i = x, y \qquad (4.42)$$

The total number of observed discreet electric fields are $2M$ and $2N$ in the x and y directions, respectively, and S_i can then be expressed as:

$$S_i = \frac{1}{2\pi}\sum_{-M}^{M}\sum_{-N}^{N} E_{Ai}(x_m, y_m) \exp\{jk_o(x \sin\theta \cos\phi + y_n \sin\theta \sin\phi)\} \Delta x\, \Delta y$$
$$x_m = m\Delta x, \quad y_n = n\Delta y \qquad (4.43)$$

In the preceding calculations, the FFT can be used to decrease the computation time. An outline of FFTs is given in Appendix B.

4.3.3 High-Power and Passive Intermodulation Tests

4.3.3.1 Definition of Passive Intermodulation

Passive intermodulation (PIM) is a technique in which a given frequency is generated by combining several frequencies that are inputs to a circuit with nonlinear input–output characteristics. This nonlinear phenomenon is observed at the junction between a connector and a transmission line when this is between different metals. The base station antenna is used for both transmission and reception and its transmitting power level is several watts per channel. The uplink receiving electric field strength is between 10 and 60 dBμV, which represents a difference from the transmitting power of around 100 dB. The received field strength is equivalent to –103 to –53 dBm in the 50Ω transmission line system. This power imbalance becomes more than 140 dB with multichannel input to base station antennas, and causes the PIM in the passive device.

Most current cellular systems are duplex, with the transmitting and receiving frequency bands allocated at a certain interval. A further receiving antenna is necessary for the base station, in addition to the antennas for transmission and reception, because of the uplink diversity system. A duplex cellular system uses the frequency bands for transmission and reception simultaneously. When the PIM occurs at the antenna for both transmission and reception, it appears on the receiving frequency band and the receiving channel cannot then be used, due to the interference at the base station.

Although the antenna is a passive device, very weak electric discharge phenomena occur at the junctions of the antenna parts during high-power operation. Such discharges often appear at the surface of the junctions when these are made of different metals. This phenomenon causes the antenna to have a nonlinear input–output response that can be regarded as a two-port circuit with transmitter input and free-space output. The output signal Y is given as the weighted summation of the input signals X by the expression:

$$Y = a_0 + a_1X + a_2X^2 + a_3X^3 + \cdots = \sum_{k=0}^{\infty} a_kX^k \tag{4.44}$$

If the two input signals have different angular frequencies, $\omega_1 = 2\pi f_1$ and $\omega_2 = 2\omega f_2$:

$$X_1 = A_1 \cos \omega_1 t, \quad X_2 = A_2 \cos(\omega_2 t + \phi) \tag{4.45}$$

where A_1 and A_2 are the respective amplitudes of the two signals and the phase angle ϕ of X_2 is normalized with respect to that of X_1. Substituting (4.45) into (4.44), the components of the nonlinear output signals can be obtained. The angular frequency components appearing at the output port can be expressed as follows, for terms up to X_i^3, for example:

1. ω_1 and ω_2 from the term for X_1;
2. $2\omega_1$, $2\omega_2$, and $\omega_1 \pm \omega_2$ from the term for X_2;
3. ω_1, ω_2, $3\omega_1$, $3\omega_2$, $2\omega_1 \pm \omega_2$, and $\omega_1 \pm 2\omega_2$ from the term for X_3.

Taking the M-signal input for X, each power series from the output of the circuit with the input–output response of (4.44) can be expressed as (4.46) by denoting each input signal as $X_i = A_i \cos\omega_i t$. Then:

$$X^k = \left(\sum_{m=1}^{M} X\right)^k = \sum_{n_1,\cdots n_M} \frac{k!}{n_1!\cdots n_M!} X_1^{n_1} X_2^{n_2} \cdots X_M^{n_M} \tag{4.46}$$

where n_i is an integer given by the expression $k = n_1 + n_2 + \cdots + n_M$, in which the sum of all the combinations of n_i is included. Then, the angular frequency ω_k of the PIM signal corresponding to this nonlinear response is given as:

$$\omega_k = n_1\omega_1 \pm n_2\omega_2 \pm \cdots n_k\omega_k \tag{4.47}$$

The parameter k is the order of PIM. The PIM signals for odd orders higher than the third order cause serious problems in current cellular systems. The interference condition for the PIM can be rewritten by denoting p as a natural number as:

$$\omega_{\text{PIM}} = \omega_2 + p(\omega_2 - \omega_1) \qquad \omega_{\text{PIM}} = p(\omega_1 - \omega_2) + \omega_1 \tag{4.48}$$

The order of PIM is $2p \pm 1$, which indicates that the odd order PIM causes the most serious problems.

Figure 4.26 shows the allocation of uplink and downlink frequency bands for cellular systems with upper and lower limit edge frequencies of f_1 and f_2, respectively. The frequency interval should be allocated so that it is

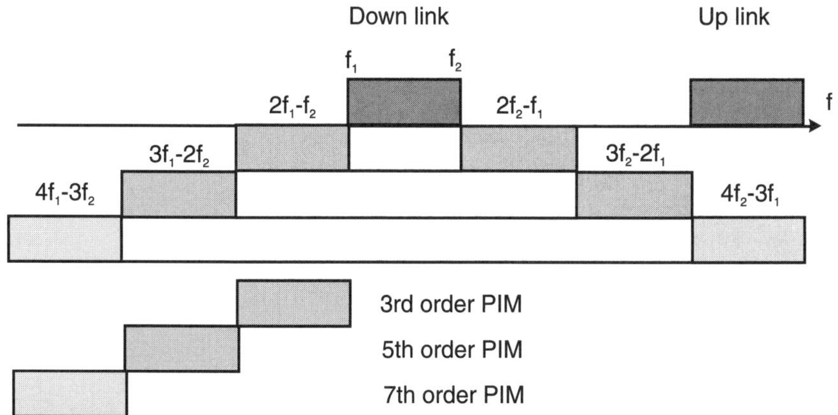

Figure 4.26 Frequency allocation for higher order PIM.

not affected by the third- and fifth-order PIM, as shown in Figure 4.26. However, the PIM should take into account the ideal frequency allocations. PIM signals are expressed as an absolute value in dBm and a relative value in dBc, normalized by the input signal. For example, for a PIM level of –100 dBm, the relative PIM value is –143 dBc, normalized by two input signals of 43 dBm.

4.3.3.2 PIM Measurement Setup

Figure 4.27 shows the block diagram of the PIM measurement setup. The measurement system consists of three units connected to the three ports of the duplexer, which operates as a transmission port, reception port, and common port for both transmission and reception. After amplification to the required power level, the two continuous signals P_1 and P_2, at different frequencies, are input to the transmitting port of the duplexer through the power combiner. The DUT is connected to the common port for both transmission and reception.

When the DUT is used as a connector and a transmission cable, the power is dissipated in a dummy load. When the DUT is an antenna, the antenna under test is set up inside an anechoic chamber, surrounded by a wave absorber to dissipate the transmitted power. With this measurement setup, the PIM signal detected at the reception port is measured by a spectrum analyzer. If the PIM signal level is less than the noise level of the spectrum analyzer, a low-noise amplifier (LNA) is inserted to adjust the receiving level. By inserting a bandpass filter prior to the LNA, the direct coupling from

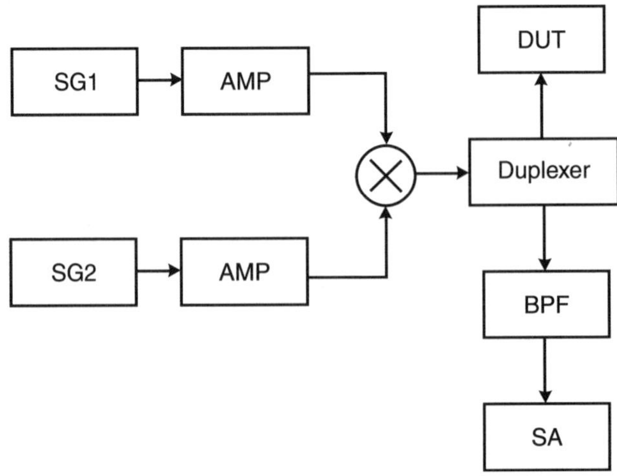

Figure 4.27 Block diagram of setup for antenna PIM measurement: SG, signal generator; AMP, amplifier; DUT, device under test; BPF, bandpass filter; SA, spectrum analyzer.

transmitting port to receiving port is eliminated. The bandpass filter suppresses the undesired PIM that occurs in the LNA and the spectrum analyzer.

When the DUT is a connector and a transmission cable, a terminating load is used, as shown in Figure 4.28. When a resistor is used as a dummy load, this also tends to have PIM within it. Therefore, a long length of coaxial cable, similar to a semirigid cable, is often used instead of a resistor. The coaxial cable of length between 50 and 100m is usually long enough to be used as the dummy load. This long cable then dissipates most of the energy. After confirming that there is no PIM excitation at the connecting lines between the duplexer and DUT, the PIM can be measured using the above setup.

During PIM measurements on antennas, metallic objects should not be located near the antenna under test in order to avoid the occurrence of PIM by induced current excitation. The PIM for antennas should preferably be measured inside an anechoic chamber, although measurements can also be made in open sites. In open site measurements, the main beam of the antenna is directed toward the sky. PIM phenomena may occur at several points, but the PIM signals can be difficult to observe, due to phase cancellation by each other. To avoid this phase cancellation, the PIM should be measured at several different frequencies.

4.3.3.3 PIM Tests for Products

PIM measurement procedures for commercial base station antennas are described in the following. Using the measurement setup described in Section 4.3.3.2, a plastic hammer is used to knock each part of the antenna under

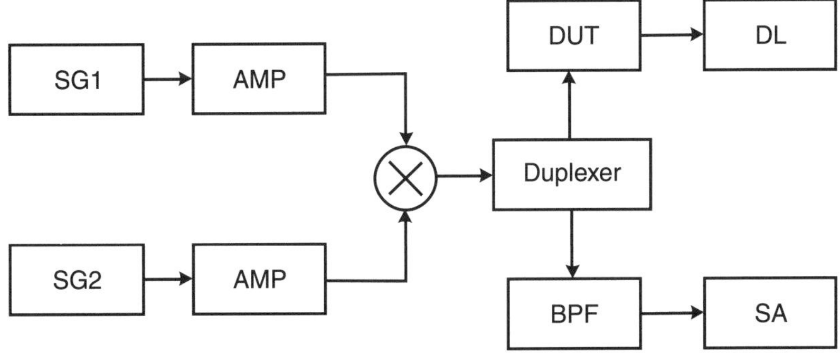

Figure 4.28 Block diagram of setup for connector and cable PIM measurement: SG, signal generator; AMP, amplifier; DUT, device under test; BPF, bandpass filter; SA, spectrum analyzer; DL, dummy load.

test. Three kinds of measured PIM phenomena can be observed: high-level PIM that occurs continuously, high-level PIM that occurs only during knocking, and variable-level PIM that occurs as a step function due to knocking. Note that the measured PIM disappears after knocking. Using this knocking test, the problem parts where the PIM occurs can be found. Note with caution, however, that the PIM phenomena sometimes reappear after a period of time.

4.4 Product Inspection

This section presents production inspection tests that should be undertaken before base station antennas are shipped. Similar tests were presented in Section 2.6 for mobile handset terminals. It is necessary for the products to satisfy specifications that have been determined in advance. In addition to the electrical performance, the input characteristics, and the radiation pattern measurements, mechanical and reliability tests (for example, vibration and moisture-proof tests) are also required with the products in use in a real environment. The antennas discussed in this section are for outdoor base stations as well as for small indoor base stations. Outdoor antennas are usually used for cellular coverage, and indoor antennas for in-building mobile communication systems or for microcellular systems. The tests presented here are not carried out on all base station antennas and in most cases, the operator and the manufacturer decide the product inspection procedures to be adopted, depending on the environment in which the antennas are intended to be used.

This section presents example specification data, but they are only samples and are based on Japanese standards. The environment in Japan is very severe for base station antennas, because there are earthquakes, typhoons, and heavy snowfalls in metropolitan areas. Inspection tests should therefore be chosen appropriately to match the environment in each country.

4.4.1 Radiation Patterns

4.4.1.1 Directivity Gain

The absolute gain (dBi) or the relative gain (dBd) of an antenna in the direction of maximum radiation is measured as the antenna gain. The measurement procedure for antenna gain is described in Section 2.2.3.1.

4.4.1.2 Radiation Pattern

The specifications for the radiation pattern for base station antennas, defined on the horizontal and vertical planes or on the E plane and H plane, are as follows: cross-polarization ratio (XPR), as defined in Section 1.2.5, and half-power beamwidth. The XPR is required to be more than 20 or 25 dB on the specified radiation plane. The beamwidth is defined by the deviation due to the frequency dependence and varying quality of the product. For example, the edge level of half-power beamwidth for 60° is 3 ± 1 dB, or 3 ± 0.5 dB at $\phi = \pm 30°$. The difference from the definition written in textbooks is due to the inclusion of the above deviation (N. Kuga, Tokyo Institute of Polytechnics, May 2000, personal communication). In addition, the polarization in the measurement is based on the slat polarization diversity antenna being measured by means of a ±45° tilted polarization antenna to coincide with the polarization for both transmission and reception. For a space diversity antenna system using only vertical polarization, the antenna pattern characteristics are tested only for vertical polarization.

4.4.1.3 Beam Tilt Angle

To adjust the coverage area of the base station, the antenna main beam is tilted an angle θ_t below the horizon. This angle is then defined as the tilt angle.

4.4.1.4 Front-to-Back Ratio

In sector zone cellular systems, the front-to-back (FB) ratio is defined as the ratio between the forward main beam level and the backward level. The definition of FB ratio is the forward radiation level at $\phi = 0°$ divided by the backward level at $\phi = 180°$. Another definition is obtained by taking the backward radiation as the maximum radiation level in the region of $120° \le \phi \le 240°$ or $90° \le \phi \le 270°$, where $\phi = 0°$ corresponds to the forward direction.

4.4.2 Antenna Input-Port Characteristics

4.4.2.1 Voltage Standing-Wave Ratio

The voltage standing-wave ratio (VSWR), defined in Section 2.1.1, at the input port of base station antennas is required to be less than 1.5 in the specified frequency band, and less than 2.0 for small indoor antennas.

4.4.2.2 Power-Resistance Capacity

For the power-resistance test, the input characteristics of the antenna should not vary for a specified input power. An example of input power might be 30W per input port. The power-resistance test is carried out at the center frequency of the specified frequency band. The procedures used in the power test should not cause any heating problems and the VSWR at the input port should not vary at high power input.

4.4.2.3 Insulation Resistance

Typically, an insulation resistance of more than 100 MΩ is required at the antenna input port, with the resistance measured by connecting the probes of an insulation resistance tester between the center conductor and the outer conductor at the input connector (M. Nakano, IDO Co., April 2000, personal communication). The maximum voltage of the tester in this example is 500V.

4.4.2.4 High Voltage-Resistance Performance

After applying 1000V AC between the center conductor and the outer conductor at the antenna input port, it should be verified that the antenna characteristics, such as VSWR, have not changed. The test procedure is the same as for the power-resistance test.

The tests for both the insulation resistance and power resistance are not applied to the antenna with a direct short circuit, as would be the case for a log-periodic antenna.

4.4.2.5 Conductivity Performance

As a conductivity performance test, the AC resistance between the inner and outer conductors at the antenna input port should be less than 100 mΩ.

4.4.2.6 Other Performance Tests

In the case of multiport base station antennas, such as polarization diversity antennas and multiband antennas, the mutual coupling between the antenna ports is defined as the isolation characteristics, as described in Section 2.1.2. The isolation should be more than 30 dB, or 25 dB in the range of specified frequency band at the two input–output ports of a polarization diversity antenna.

4.4.3 Feed Circuit

Recently developed base station antennas adjust their coverage areas by changing their tilt angle in the vertical plane, and both mechanical and electrical methods can be used for beam tilting. In the electrical method, the base station antenna consists of several subarray units, and the phase of each is electrically changed for beam tilting. Therefore, the antenna system has two

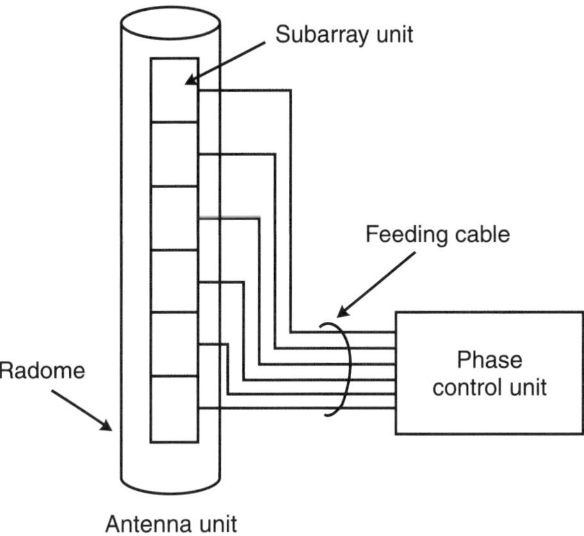

Figure 4.29 Phase control unit and base station antenna.

components, these being the antenna itself and a feed circuit for phase adjustment. Three types of phase adjusting system can be used for the feed circuit: remote control, manual control, and fixed advance.

The performances of the control schemes mentioned are similar and one or the other will be selected, depending on the cost of the base station(s). The phase control unit is located separately from the antenna unit, as shown in Figure 4.29. This feed circuit is tested independently of the antenna unit. The measurements made are the same as for antenna measurements, and include VSWR, power-resistance capability, insulation resistance, high voltage-resistance performance, conductivity performance, and mutual coupling between input ports. These tests are not made on antennas operated at a fixed beam tilt angle, and those in which phase differences are provided to each antenna element.

4.4.4 Appearance and Weight

The dimensions and total weight of each antenna part should be within specification. The nameplate, logo, and painting color should also be confirmed.

4.4.4.1 Antenna Construction

The antenna construction, as shown in the specifications, must be verified.

4.4.4.2 Assembly Instructions

The assembly instructions are required to be sufficiently comprehensive such that mistakes will not be made during installation.

4.4.4.3 Range of Mechanical Adjustment

If the beam tilt adjustment is performed mechanically, the range of adjustment available must be verified. In addition, the adjustment range of the movable polarization plane should also be checked, if applicable.

4.4.4.4 Wind Pressure Resistance

Depending on the wind pressure on the antenna, it should be confirmed that the radome and the other antenna parts have sufficient mechanical strength to withstand a wind velocity v up to 60 m/s. This check is usually carried out by numerical calculation. If the wind pressure is q (kgf/m^2) and the wind pressure coefficient is C, the wind pressure loading P (kgf/m^2) is:

$$P = Cq \tag{4.49}$$

where

$$q = 120\sqrt[4]{h} \quad (h > 16\text{m}), \qquad q = 60\sqrt[4]{h} \quad (h \le 16\text{m}) \tag{4.50}$$

and h is the antenna height. The wind pressure coefficient C for antenna pillars with square and circular cross-sections has values of 2 and 0.8, respectively. Therefore, a circular column is preferable from the viewpoint of wind pressure. In a typical example, the wind pressure loading can be larger than 350 kgf/m^2.

In addition, the allowed rolling vibration range is within ±2 degrees for a wind velocity of 60 m/s. Furthermore, no noise should be generated by the wind.

4.4.4.5 Other Tests

If a phase control unit is added to the antenna system, its tolerance should also be checked.

4.4.5 Reliability Examination

The reliability of antennas is examined according to the standard regulations of the International Electrotechnical Commission (IEC) and typical tests involved are described in this section.

4.4.5.1 Thermal Test (Heat Cycle Test)

The antenna characteristics are examined in a heat cycle test under conditions in which the humidity is more than 90% and the temperature range is from –30°C to +60°C. The actual heat cycle is as follows: After the temperature has been raised from a room temperature of +20°C, the antenna is held at the highest temperature of +60°C for several hours. The temperature is then dropped to the lowest level of –30°C and is held at this level for several more hours. It is then raised again to the highest temperature. This procedure represents one cycle of the thermal test. During a complete test program, the input VSWR is checked after tens of cycles of tests. For small indoor base station antennas that can be tested inside a compact box, the variations of temperature with time during a heat cycle test are as described below (H. Sugisawa, NHK Spring Co., Ltd., April 2000, personal communication). The temperature is raised from the room level of +20°C to +50°C in 1.5 hours, and this high temperature level is held for 3 hours. The temperature is then dropped from +50°C to –20°C in 3 hours, and the temperature held at –20°C for 3 hours. Finally, the temperature is raised again to +20°C in 1.5 hours. This variation corresponds to one heat cycle.

When automatic temperature control is not available, the following is used as one heat cycle: The temperature is held at +50°C for 4 hours after raising it from room level, then a temperature of –20°C is maintained for 4 hours. Finally, the antenna under test is left at room temperature for a further 2 hours. The antenna characteristics, such as the input VSWR, are evaluated several times during the heat cycle.

4.4.5.2 Heat (Cold) Resistance Performance

Vehicular-mounted antennas are tested for a few hours in a heat-resistance test at 90°C and a cold-resistance test at –40°C. As a heat shock test, the heat/cold-resistance test covers a period of 3 hours and is repeated several times. However, this test is not often performed.

4.4.5.3 Waterproofing and Moisture-Proof Performance

The waterproofing test is carried out on the product before shipping. The moisture-proof test is applicable to both the parts and the materials of the antennas. An example of a waterproofing test is as follows: In the same installation as intended for real use, water is sprinkled all over the antenna using a nozzle mounted above the antenna under test at a height of from 300 to 500 mm. As an example of the sprinkling rate, the quantity of water can be 10 ± 0.5 liters per minute, with a pressure of between 50 and 150 kPa for more than 1 hour. On completion of this waterproofing test, it is confirmed that the insulation resistance at the antenna input port has not changed.

A sealed-type antenna used for small base stations can be immersed at the bottom of a water tank at a depth of more than 400 mm for 24 hours, and the penetration of water is then confirmed by breaking up the antenna. The moisture-proof performance of small base station antennas can be tested using a box at constant temperature and humidity. As an example of a typical test cycle, the antenna is left inside the box at a temperature of 40°C and a humidity of 80% for 96 hours. It is then placed in an area at room tem-

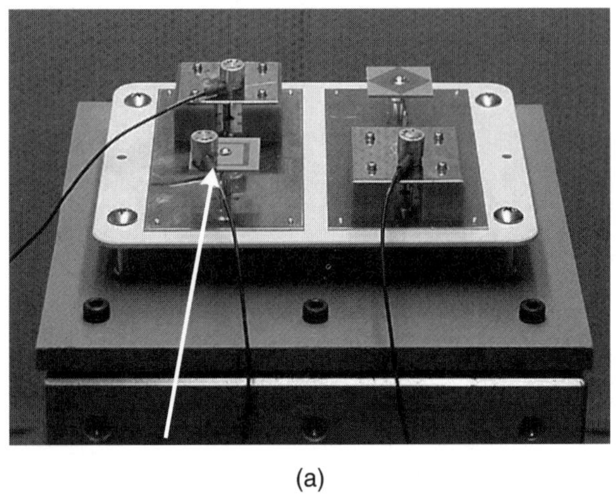

(a)

(b)

Figure 4.30 Vibration test: (a) probe attachment for vibration test and (b) vibration test setup.

perature and humidity for 1 hour. After this cycle, the antenna input VSWR characteristics should not have changed.

Note that these waterproof and moisture-proof tests are applied only to a sample of the products (T. Matsuoka, Nihon Dengyo Kousaku Co., Ltd., April 2000, personal communication). The tests carried out before shipping include VSWR measurement and appearance checking.

4.4.5.4 Vibration Test

For antennas used in areas subject to vibration, such as on highways, a vibration test is required. Before the vibration test, the mechanical resonance frequency is measured by vibrating the antenna at variable frequency in both the vertical and horizontal directions, with an accelerometer attached to each part of the antenna. An example measurement setup is shown in Figure 4.30.

The antenna is then vibrated at the measured resonant frequency in the x, y, and z directions. In a typical test example, a force of 5G is applied for 1 hour in the z direction and for 30 minutes in the x and y directions. After the vibration test, it must be confirmed that the VSWR has not changed.

References

[1] "Propagation Data and Prediction Methods for the Terrestrial Land Mobile Service Using Frequency Range 30 MHz to 3 GHz," CCIR SG-5, Report 567-3 (MOD F).

[2] Kozono, S., and K. Watanabe, "Influence of Environmental Buildings on UHF Band Mobile Radio Propagation," *IEEE Trans. Commun.*, Vol. COM-25, No. 10, Oct. 1977, pp. 1133–1143.

[3] Hata, M., "Empirical Formula for Propagation Loss in Land Mobile Radio Service," *IEEE Trans. Vehicular Technology*, Vol. VT-29, No. 3, 1980, pp. 317–325.

[4] Green, E., and M. Hata, "Microcellular Propagation Measurements in an Urban Environment," *Proc. 1991 IEEE Int. Symp. on Indoor and Mobile Radio Communications,* King's College, London, Sept. 1991, pp. 23–25.

[5] Blackar, K. L., et al., "Path Loss and Delay Spread Models as Functions of Antenna Height for Microcellular System Design," *Proc. 1992 IEEE Vehicular Technology Conf.*, Denver, CO, May 1991, pp. 333–337.

[6] Oda, Y., T. Tanaka, and K. Sato, "Microwave Band LOS Path Loss Characteristics in Microcellular Mobile Communications," *Proc. Int. Symp. on Antennas and Propagation, ISAP'96*, Chiba, Japan, Sept. 1996, pp. 1097–1100.

[7] Oda, Y., and K. Tsunekawa, "Advanced LOS Path Loss Model in Microwave Mobile Communications," *ICAP'97*, Apr. 1997, pp. 2.170–2.173.

[8] Fujimori, K., and H. Arai, "Indoor Propagation Characteristic Including Radiation Pattern and Polarization of Base Station," *Proc. IEEE AP-S Int. Symp. Conf. Digest,* Montreal, Canada, July 1997, pp. 2006–2009.

[9] Taylor, T. T., "Design of Line-Source Antennas for Narrow Beamwidth and Low Sidelobe," *IRE Trans. Antennas and Propagation,* Vol. AP-3, Jan 1955, pp. 16–28.

[10] Ebine, Y., "Dual Frequency Base Station Antennas for PDC Systems in Japan," *Proc. IEEE AP-S. Int. Symp. and USNC/URSI National Radio Science Meeting*, Orlando, FL, July 1999, pp. 564–567.

[11] Fujii, T., et al., "Handling Capacity Expansion of Land Mobile Communications," *NTT Review*, Vol. 2, No. 3, Mar. 1990, pp. 55–61.

[12] Yamada, Y., Y. Ebine, and K. Tsunekawa, "Base and Mobile Station Antennas in Land Mobile Radio Systems," *Trans. IEICE Japan,* Vol. E74, No. 10, June 1991, pp. 3202–3209.

[13] Taga, T., and K. Tsunekawa, "A Built-In Antenna for 800 MHz Band Portable Radio Units," *Proc. ISAP'85,* Kyoto, Japan, 1985, pp. 425–428.

[14] Vaughan, R., "Polarization Diversity in Mobile Communications," *IEEE Trans. Vehicular Technology*, Vol. VT-39, No. 3, Aug. 1990, pp. 177–186.

[15] Turkmani, A. M. D., et al., "An Experimental Evolution of the Performance of Two Branch Space and Polarization Diversity Schemes at 1800 MHz," *IEEE Trans. Vehicular Technology,* Vol. VT-44, No. 2, May 1995, pp. 318–326.

[16] Lotse, F., et al., "Base Station Polarization Diversity Reception in Macrocellular System at 1800 MHz," *Proc. IEEE VTS 46th Vehicular Technology Conf.*, Apr. 28–May 1, 1996, Vol. 3, pp. 1643–1646.

[17] Arai, H., and M. Nakano, "Up-Link Polarization Diversity and Antenna Gain Measurement of Handheld Terminal at 900 MHz," *Proc. MDMC'94*, Niigata, Japan, Nov. 21–22, 1994.

[18] Yaghjian, A. D., "An Overview of Near-Field Antenna Measurement," *IEEE Trans. Antennas and Propagation*, Vol. AP-34, No. 1, 1986, pp. 30–45.

[19] Kong, J., *Electromagnetic Wave Theory,* 2nd ed., New York: John Wiley & Sons, 1990, pp. 307–312.

5

Fading and Field Simulators

In mobile communication systems, modulated electromagnetic waves arrive at a reception point as a summation of several waves that have been propagated along different paths. The characteristics of these waves are distributed randomly, and this causes spatially distributed standing waves. Under these conditions, when the receiving position moves, the received electric field strength varies greatly. This phenomenon is called *interference fading* or *multipath fading*, and is the most serious problem in mobile communication systems.

Mobile terminals are tested at the development stage to ascertain whether or not they can satisfy the specifications when fading is present. The tests involve propagation measurements in a real outdoor environment. The random field measurement described in Chapter 1 is a well-known method for testing the performance of mobile terminals in a multipath propagation environment. However, the outdoor propagation measurement is not always stable in time but is necessary in order to obtain a license and hence satisfy the regulations. These considerations indicate the need for an indoor facility to reproduce the real multipath propagation environment.

Two methods can be used to reproduce fading phenomena. One method generates fading waves as a radio-frequency (RF) output signal from a radio unit following the antenna port. Such an arrangement constitutes a fading simulator. The other method reproduces fading waves as a summation of spatially distributed standing waves in an artificial multipath propagation environment, typically in an indoor facility. This is another type of field simulator in which radio equipment products can be tested.

This chapter describes the principles and operation of fading and field simulators, which are indispensable for the development of mobile terminals.

5.1 Fading Simulators

The reception voltage at the input port of a radio unit can be regarded as a function of time for moving mobile communication terminals, because the RF signal transmitted from the base station arrives at the reception point as a summation of multipath waves and excited standing waves. The reception voltage often falls to a level close to the noise level, which then constitutes a fading phenomenon. In the development of communication systems and radio equipment, the performance should be confirmed under real fading conditions. However, real propagation measurements have the problems described above and, therefore, the use of fading simulators is mandatory for tests on radio equipment for mobile communication systems. The required performance of a fading simulator must include the reproduction of Doppler fading due to the velocity of mobile terminals, more than one correlated reception signal with fading, and wideband characteristics to cover the allocated frequency bands. This section presents the theory and circuit structure of such a fading simulator.

5.1.1 Theory of Fading Simulators

We first discuss the theory behind fading simulators [1, 2]. In mobile communication cellular systems, a mobile terminal moving in a multipath fading environment receives an electric field with a varying amplitude due to Rayleigh distribution and with randomly distributed phase [3]. This fading signal can be expressed, as shown in Section 1.2.3, by assuming that a large number of waves arriving at a reception point have almost the same amplitude, as follows:

$$E_r(t) = x(t)\cos\omega_c t - y(t)\sin\omega_c t \tag{5.1}$$

where

$$x(t) = \frac{1}{\sqrt{2\pi\sigma}}\exp\left(-\frac{x^2}{2\sigma^2}\right), \quad y(t) = \frac{1}{\sqrt{2\pi\sigma}}\exp\left(-\frac{y^2}{2\sigma^2}\right) \tag{5.2}$$

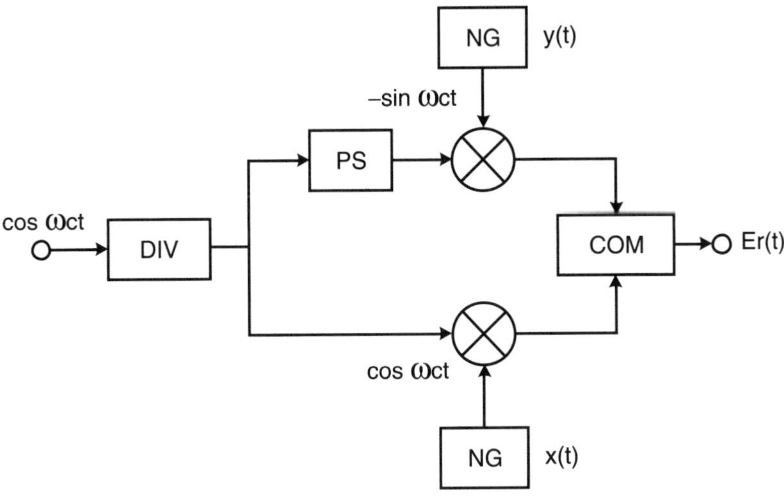

Figure 5.1 Block diagram of fading simulator: DIV, power divider; PS, phase shifter; NG, Gaussian noise generator; COM, power combiner.

The preceding equations indicate that fading signals can be obtained by modulating the in-phase component cos $\omega_c t$ and the orthogonal component sin $\omega_c t$ with two independent baseband noise signals $x(t)$, $y(t)$, respectively, as shown in Figure 5.1. In this case, the modulation is double sideband with carrier suppression. The power spectra for the two baseband noise signals used for the modulation are given as functions of the velocity of the mobile terminal v (m/s) and the maximum Doppler frequency $f_D = v/\lambda$ (Hz) as:

$$S_x(f) = \begin{cases} S_y(f) = \dfrac{\sigma^2}{\pi f_D} \dfrac{1}{\sqrt{1-(f/f_D)^2}} & \text{for } 0 \le f \le f_D \quad (5.3a) \\ 0 & \text{for } f_D \le f \quad (5.3b) \end{cases}$$

The above Gaussian noise levels at the baseband frequency are obtained by restricting the frequency band width of the M-sequence code input to a low-pass filter with the same transmission function as above, that is:

$$f_D = \frac{f_b}{20} \quad (5.4)$$

Although a cycle of the output signal of the fading simulator is the same as that of the M-sequence code, an effective length of more than 1,600 wavelengths is sufficient for the measurement facility. The M-sequence (maximum length code) is used to obtain random sequences. The M-sequence generator can be easily constructed from a shift register and an exclusive OR (XOR) operator, an arrangement that is often employed in digital communication systems to synchronize input and output signals.

5.1.2 Fading Simulator for Correlated Signal Outputs

One technique that reduces fading effects in mobile communication systems is diversity reception with more than one input branch and small correlation coefficients. To test diversity reception, it is necessary for the fading simulator to incorporate the functions that make the correlated output signals. Assume the two fading signals with correlation to be the following:

$$E_1(t) = x_1(t)\cos\omega_c t - y_1(t)\sin\omega_c t \tag{5.5a}$$

$$E_2(t) = x_2(t)\cos\omega_c t - y_2(t)\sin\omega_c t \tag{5.5b}$$

where the mean values of both $x_i(t)$ and $y_i(t)$ are zero, they have the same variance σ and Gaussian distributions given by:

$$x_1(t) = u_1(t)\cos\phi_1, \quad y_1(t) = v_1(t)\sin\phi_1 \tag{5.6a}$$

$$x_2(t) = u_2(t)\cos\phi_2, \quad y_2(t) = v_2(t)\sin\phi_2 \tag{5.6b}$$

In the preceding equations, the following two combinations have correlation and the others have no correlation:

$$\langle u_1(t)\cdot u_2(t)\rangle = \langle v_1(t)\cdot v_2(t)\rangle = \sigma^2\cos(\phi_1 - \phi_2) \tag{5.7a}$$

$$\langle u_1(t)\cdot v_1(t)\rangle = \langle u_1(t)\cdot v_2(t)\rangle = \langle u_2(t)\cdot v_1(t)\rangle = \langle u_2(t)\cdot v_2(t)\rangle = 0 \tag{5.7b}$$

where $<x>$ denotes the mean value of the variable x, and the combined probability function of the above is:

$$p(u_1, u_2, v_1, v_2) = \frac{1}{(2\pi\sigma^2)(1-\rho^2)}$$

$$\exp\left(-\frac{1}{2\sigma^2(1-\rho^2)}(u_1^2 + u_2^2 + v_1^2 + v_2^2 - 2\rho u_1 u_2 - 2\rho v_1 v_2)\right) \quad (5.8)$$

where $\rho = \cos(\phi_1 - \phi_2)$.

The Gaussian distributions can be replaced by the following probability density function using the same procedure as for (1.7) and (1.8) in Chapter 1:

$$p(R_1, R_2) = \frac{R_1 R_2}{\sigma^4(1-\rho^2)} \exp\left(-\frac{R_1^2 + R_2^2}{2\sigma^2(1-\rho^2)}\right) I_0\left\{\frac{\rho R_1 R_2}{\sigma^2(1-\rho^2)}\right\} \quad (5.9)$$

where I_0 is a modified zero-order Bessel function and the correlation coefficient ρ_R of R_1 and R_2 is defined as:

$$\rho_R = \frac{\langle (R_1 - \langle R_1 \rangle)^* (R_2 - \langle R_2 \rangle) \rangle}{\sqrt{\langle (R_1 - \langle R_1 \rangle)^2 (R_2 - \langle R_2 \rangle)^2 \rangle}} \cong \rho^2 = \cos^2(\phi_1 - \phi_2) \quad (5.10)$$

As a result, two correlated fading signals are generated by modulating the inphase and orthogonal components with four independent baseband noise signals, after linear transformation of (5.6). The corresponding circuit diagram is shown in Figure 5.2. The electrical circuit required to generate Gaussian noise signals $u_1(t)$, $v_1(t)$ from four independent Gaussian noise signals consists of the operational amplifier and resistance shown in Figure 5.3. The correlation coefficient is arbitrarily defined by ϕ_1 and ϕ_2.

5.1.3 Delay Spread Simulation

The fading simulator described in the previous section generates Rayleigh fading at a single frequency. In mobile communication systems, any modulated RF signal has its own frequency bandwidth, which is affected by frequency-selective fading. The propagation path from a transmitting source to a reception point has a different length for each frequency, and causes delay spread distribution, especially in high-speed digital data transmissions. To

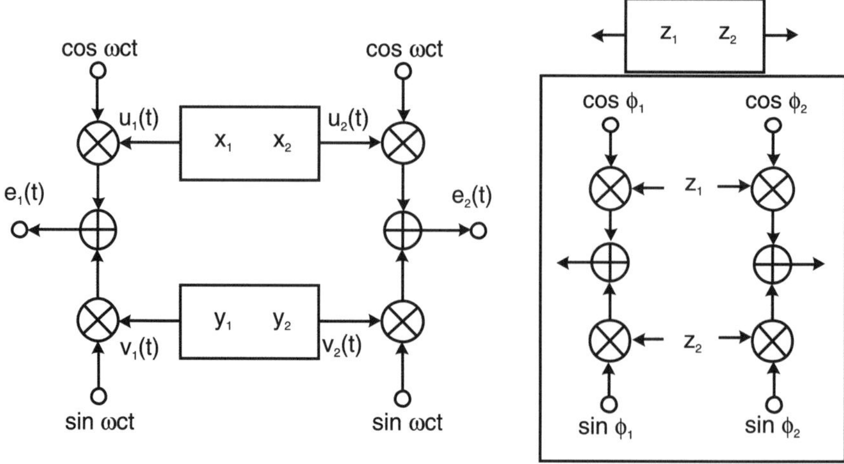

Figure 5.2 Block diagram for generation of two correlated fading waves.

examine mobile terminals under this delay spread condition, a fading simulator is required to have frequency-selective fading characteristics.

A block diagram for a fading simulator that can reproduce this type of frequency selective fading is shown in Figure 5.4. In Figure 5.4, split input RF signals derived from a power divider pass through each delay circuit to give different delay times. The delay circuits consist simply of coaxial cables with different lengths, and the variable delay times are obtained using pin diode switches and microstrip lines. The amplitude of each delayed signal is adjusted by multiplication by an arbitrary coefficient, and all the delayed signals are combined in the last stage by the power combiner [4].

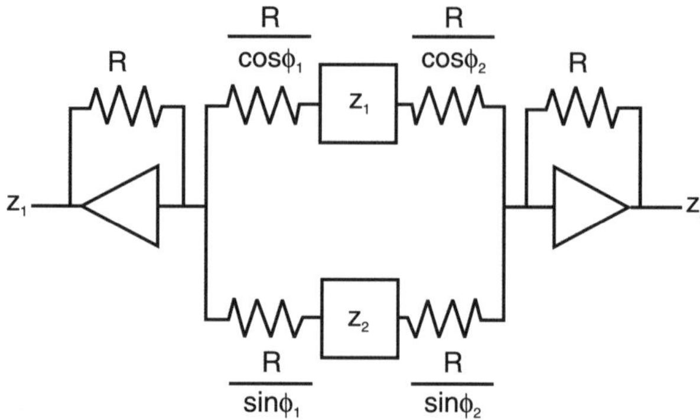

Figure 5.3 Block diagram for generation of correlated Gaussian noise.

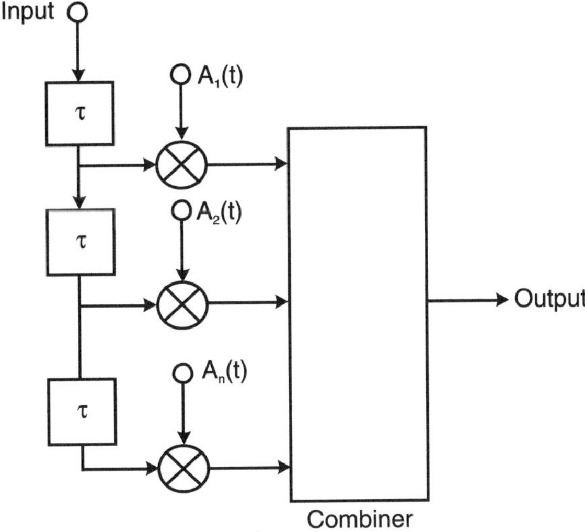

Figure 5.4 Block diagram for delay spread simulator, where τ denotes a delay circuit.

For frequency-selective fading in a real propagation environment, the delay time, the amplitude, and the phase of the delayed wave all behave as functions of time. Measured delay times for current digital cellular systems vary from a few hundreds of nanoseconds (ns) to tens of microseconds (μs). To obtain real frequency-selective fading, the transverse filter implemented by LSI in Figure 5.5 must process the signal in the time domain. This signal processing fills the baseband frequency with digital data, which gives detailed parameter settings for delay spread simulation. The synthesized digital signal is input to a bandpass filter after digital-to-analog (D/A) conversion, and is modulated by Gaussian noise to generate fading waves [5]. This type of field simulator is used to evaluate the equalizer used to suppress the delayed waves in mobile terminals.

5.1.4 Bit Error Ratio Measurement Using a Fading Simulator

A bit error ratio (BER) is used to evaluate the diversity performance of handset terminals in digital cellular systems in a multipath propagation environment. The number of error codes calculates the BER by inputting the correct codes in advance of testing the equipment. Figure 5.6 shows a block diagram of a system for evaluating diversity performance [6]. By setting the measurement frequency as the center frequency of the RF signal generator, the median value of the detected signal level, observed by the power meter, is adjusted to a specified value by the attenuator. After this setting has been made, the

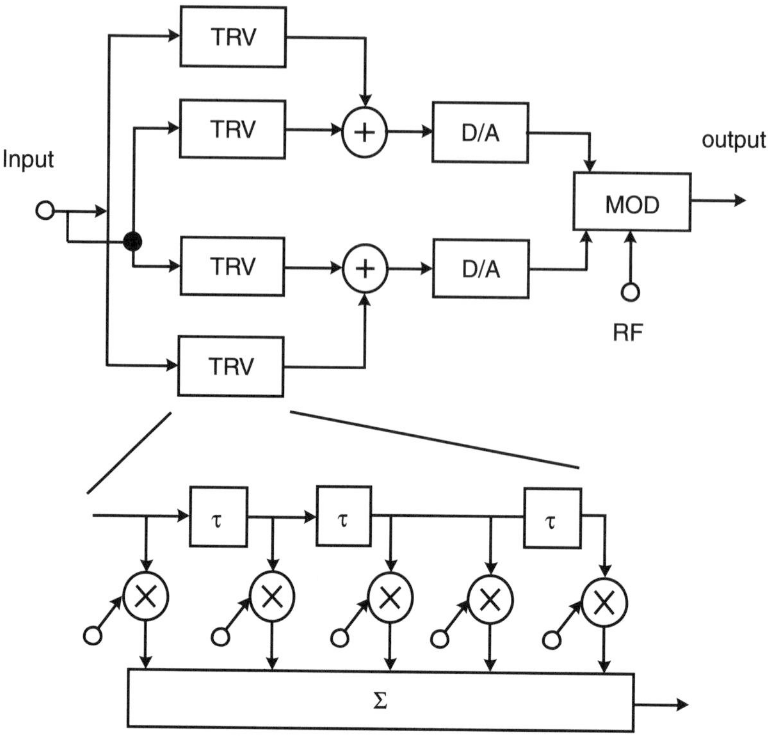

Figure 5.5 Block diagram for real-time delay spread simulator: TRV, transversal filter; D/A, digital-to-analog converter; MOD, modulator; τ, delay circuit.

output port is switched to the equipment under test and the BER is measured. The maximum BER is specified as a standard for handset terminals. In diversity performance evaluation, the levels of two output ports are adjusted using the same procedure as that described above and the BER is measured by connecting two output ports to the test equipment.

5.2 Field Simulators

The recent remarkable increase in the number of subscribers in mobile communications requires miniaturized and high-performance handset terminals. Using the fading simulator described in the preceding section, the characteristics of handset terminals can be evaluated by inputting a fading wave to the antenna input port with the handset antenna removed. The test is effective for evaluating the radio unit of a handset terminal without the antenna.

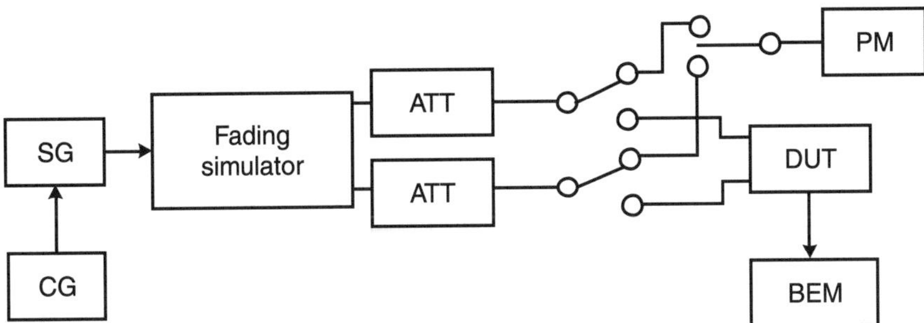

Figure 5.6 Block diagram for BER measurement using fading simulator: CG, standard code generator; SG, signal generator; ATT, attenuator; DUT, device under test; PM, power meter; BEM, bit error meter.

However, it is also necessary to confirm the characteristics of the product with the antenna included. The product test performed inside an anechoic chamber, as described in Chapter 2, is not sufficient for mobile terminals intended for use in urban multipath propagation environments. Although the random field measurement is effective in confirming the characteristics of the handset, the stable measurement environment present in an indoor facility is required for the development of handset terminals.

This section presents the theory and structure of a field simulator for the evaluation of handset terminals in an artificial multipath propagation environment. The fading simulator reproduces fading signals as RF signals fed through a coaxial cable, whereas the field simulator generates fading waves as a summation of spatially distributed electromagnetic waves.

By installing multiple radiating sources and scattering obstacles inside a large anechoic chamber, artificial fading can be generated as a field simulator. However, such a large facility is not appropriate in the design process for handsets, and a more compact field simulator is required that can be set up in a laboratory. The following material in this chapter describes the theory and structure of a field simulator for handset evaluation in artificial multipath propagation environments. In addition, measurements on the diversity antenna built in to a handset terminal are included.

5.2.1 Theory of Field Simulators

Simple fading can be measured by moving a device under test (DUT) along a standing wave generated by a feed probe and reflectors, as shown in Figure 5.7. The feed probe is constructed from a monopole antenna installed on a large ground plane, and is placed a quarter wavelength ($\lambda/4$) away from one

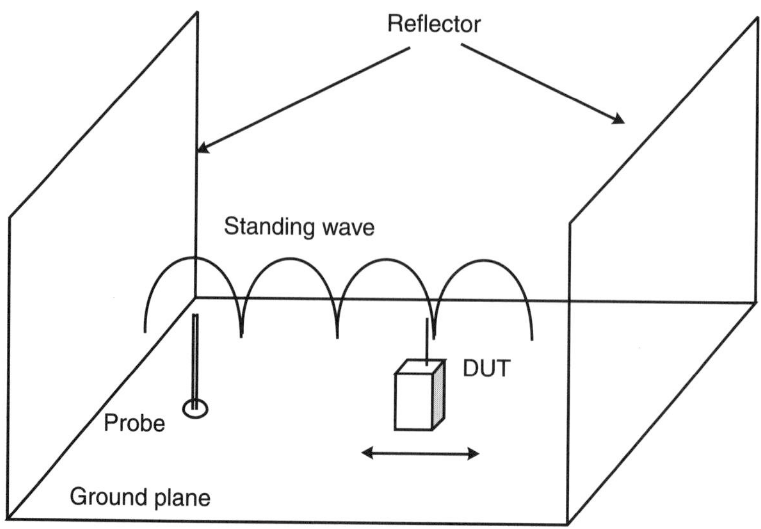

Figure 5.7 Fading measurement using standing waves.

reflector in order to excite the standing wave effectively. The distance between the two reflectors is set to $m\lambda/2$, where m is a natural number [7].

The number of incoming waves to a mobile terminal in a cellular system is seven or eight [8], and this is quite different from the standing-wave case described earlier, which consists of only two waves. Several transmitting antennas installed inside the anechoic chamber can reproduce the same multipath fading environment, and then the characteristics of the DUT mounted on a rotator inside a chamber can be measured under fading conditions. A disadvantage of this system is that a large amount of space is required.

In an actual solution, the multipath fading generated by seven or eight waves is reproduced in the limited space, and the fading waves at an observation point can be written as a function of time. To generate interference fading using a limited number of sources, the reflected waves from the conducting reflector must be appropriate for the field simulator. A compact field simulator consists of the transmitting antennas installed on the walls inside a shielded room. In the following section, this type of field simulator is described in detail.

5.2.2 Fundamentals of Field Simulators

The basic parameters for field simulators are the number of sources and the dimensions of the box. The required performance involves the generation of arbitrary probability distributions from Rayleigh to Rician, and the arbitrary correlation coefficient between two observation points. Frequency shifting due to the Doppler effect of the moving terminal is not considered for simplicity of the simulator structure.

The field simulator, surrounded by conducting walls, is regarded as an oversized rectangular waveguide terminated by shorting boards at both ends, as shown in Figure 5.8. Assuming that a standard dipole antenna is installed inside a waveguide as the source which excites the electric fields, the field distribution inside the simulator is solved by expanding modes existing in the waveguide [9]. With the field distribution regarded as a function of time, the phase of each source is changed electrically at random by a digital phase shifter. An eight-bit digital phase shifter is used in the following example in order to satisfy the requirement that the minimum number of bits should be more than five [10]. The random phase variation means that the radiated field from the sources is scanned electrically. This is a type of phased-array antenna, which moves the electric field distribution at the reception point at very high speed.

The minimum number of sources is an important parameter for the field simulator. Although a single source allows a standing wave to be generated inside a simulator using the reflection waves from the walls, fields varying as functions of time require more than one source. The random phase variation of two sources produces Rayleigh or Rician fading at the observation point, depending on the cumulative probability distribution inside the simulator. As shown in Figure 5.9, the calculated probability distributions

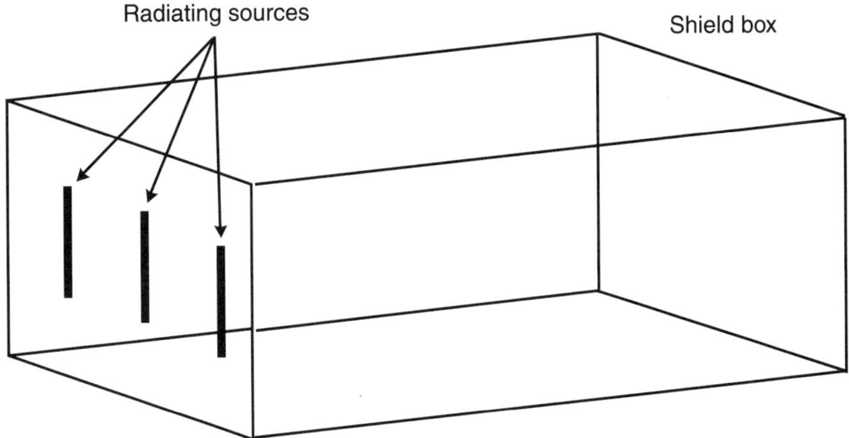

Figure 5.8 Field simulator using shield box.

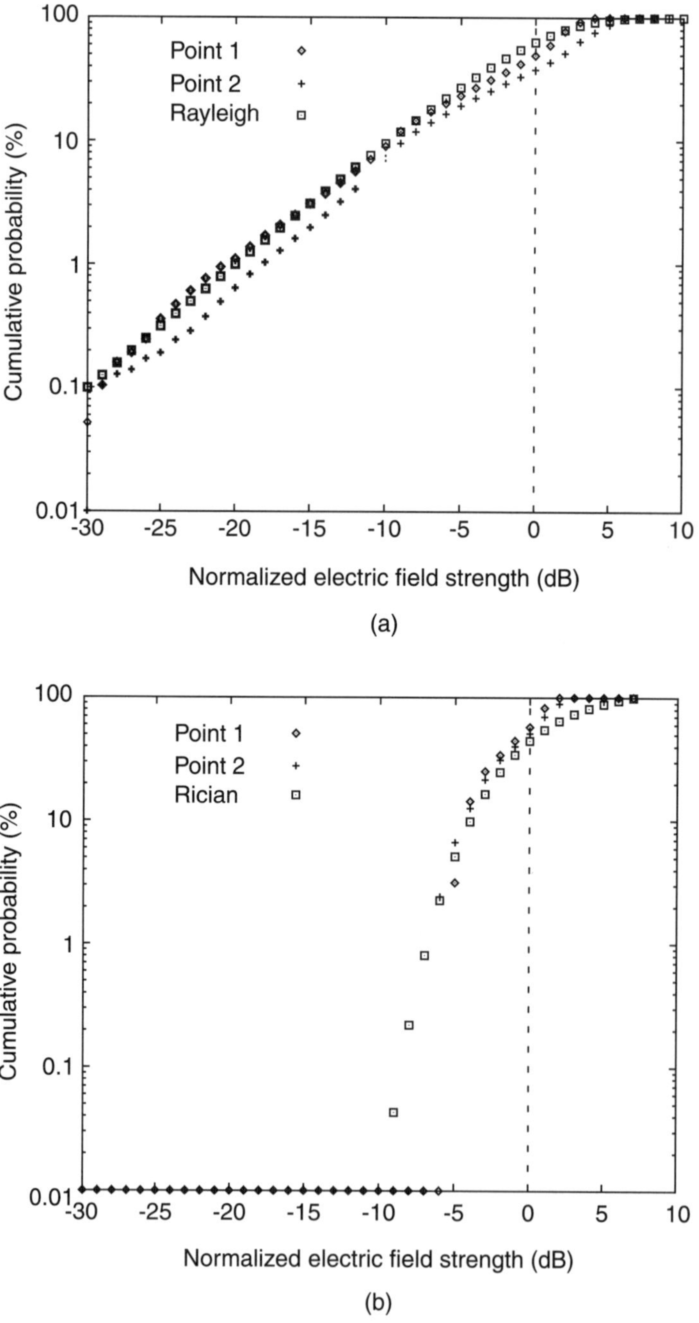

Figure 5.9 Cumulative probability distribution: (a) Rayleigh position with two sources and (b) Rician position with two sources.

become nearly equal to Rayleigh and Rician curves, which indicates that two sources are adequate for the point of field distribution.

However, the correlation coefficient between two observation points does not become small enough for diversity measurement. The correlation coefficient of the two-source case is 0.8 when the distance between observation points is 3.5λ. The correlation coefficient for two vertically orientated antennas at a mobile terminal in a cellular system is given here by (5.11), with the assumption of uniform distribution of incoming waves in the horizontal plane [11]:

$$\rho_e = J_0\left(\frac{2\pi d}{\lambda}\right)^2 \tag{5.11}$$

where d is the distance between two observation points, and $J_0(x)$ is the zero-order Bessel function. The small correlation coefficient is obtained using three sources. Figure 5.10(a) shows the corresponding calculated histogram of correlation coefficients. The value $\rho_e = 0.1$ is almost the same as that calculated using (5.11), and its histogram is quite similar to the distribution measured using random field measurement (RFM) described in Section 2.5.3, as shown in Figure 5.10(b). The value $\rho_e = 0.1$ is calculated from the average of this histogram.

Although size can be regarded as one of the parameters of the field simulator, the number of sources is the dominant factor that characterizes its performance. The size of the simulator is determined by the size of the device under test.

5.2.3 Hardware Components for Field Simulators

This section presents examples of hardware components for field simulators. The field distribution can be evaluated by the mode expansion technique with a rectangular waveguide, as described in the preceding section. However, ray-tracing analysis is introduced in this section as a means of following the arbitrary values of the reflection coefficient from simulator walls [12].

The field simulator described in this section is shown in Figure 5.11. The simulator is 1.65m wide, 1.33m high, and 2m deep. These are minimum values for the evaluation of handset terminals in the 800-MHz band. These sizes are equivalent to 4.51λ × 3.64λ × 5.47λ at the measurement frequency of 820 MHz. Figure 5.12 shows the general form of the field simulator.

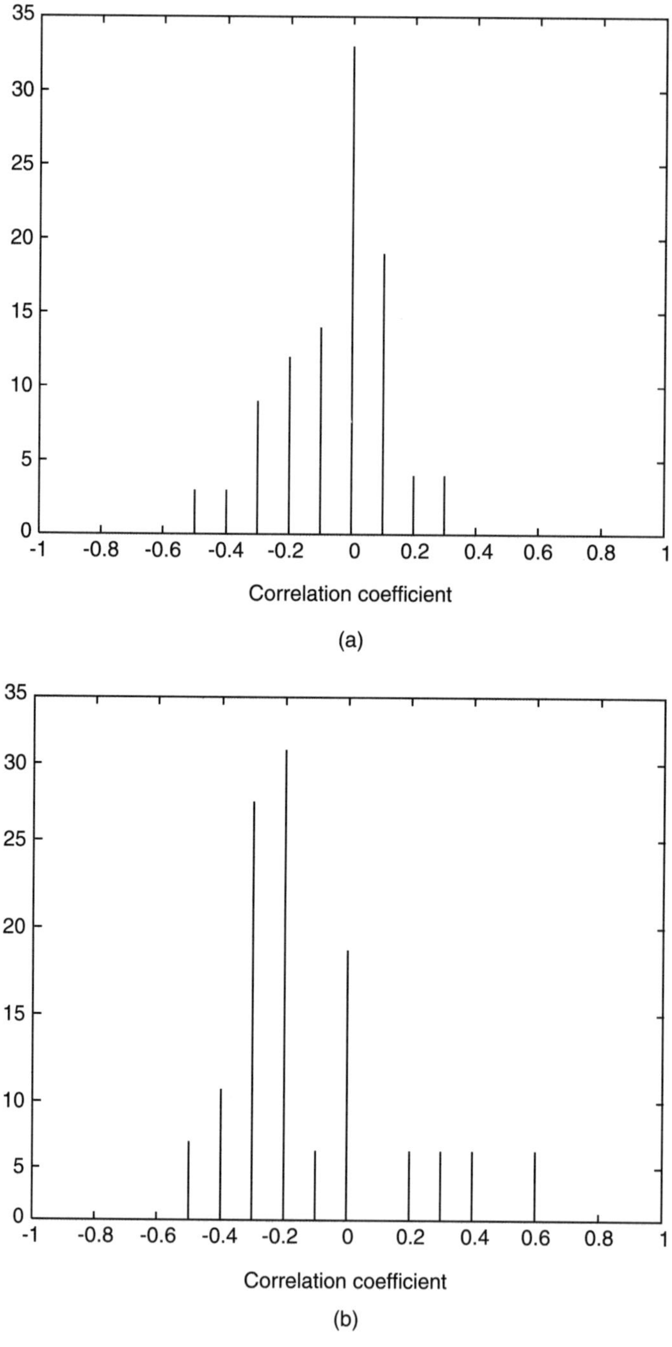

Figure 5.10 Histogram of correlation coefficient: (a) three-source case and (b) measured results using RFM.

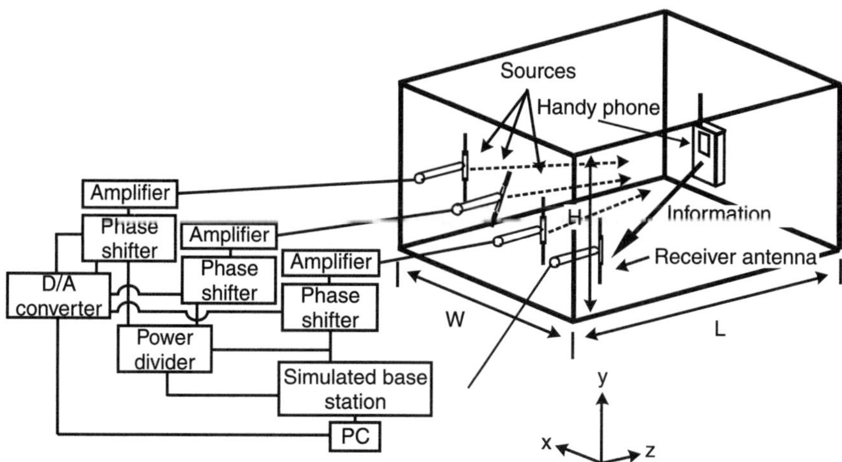

Figure 5.11 Structure of a field simulator.

The wall opposite the wall where the sources are mounted is covered by a wave-absorbing material, and the rest of the walls are covered with conducting metal. The handset under test is located near the absorber wall so as to avoid the input impedance change of the test terminal. Three standard dipole antennas are mounted at the offset positions so as to excite Rayleigh and Rician field distributions in an effective manner. A signal tester, controlled by a personal computer (PC), controls the sources in such a way that their phases are randomly changed to excite fading waves, after passing through a three-way power divider.

The handset under evaluation is fixed at a specific position, and the fading pitch is varied by changing the switching speed of the phase shifter. This control of speed enables the fading pitch to be changed and provides various different propagation environments. In this particular field simulator configuration, no cable is connected to the handset to eliminate any effects the cable might have on the measurement. The received field strength at the handset is converted to the intermediate frequency (IF), and the output voltage level of the received signal strength indicator (RSSI) is transmitted from the handset. An additional reception-only dipole antenna picks up this signal, and reports the received electric field strength of the handset under test.

5.2.3.1 Electric Field Distributions

As shown in Figure 5.13, the electric field distributions inside the simulator are calculated using the formulas defined next. The wave absorber covers one

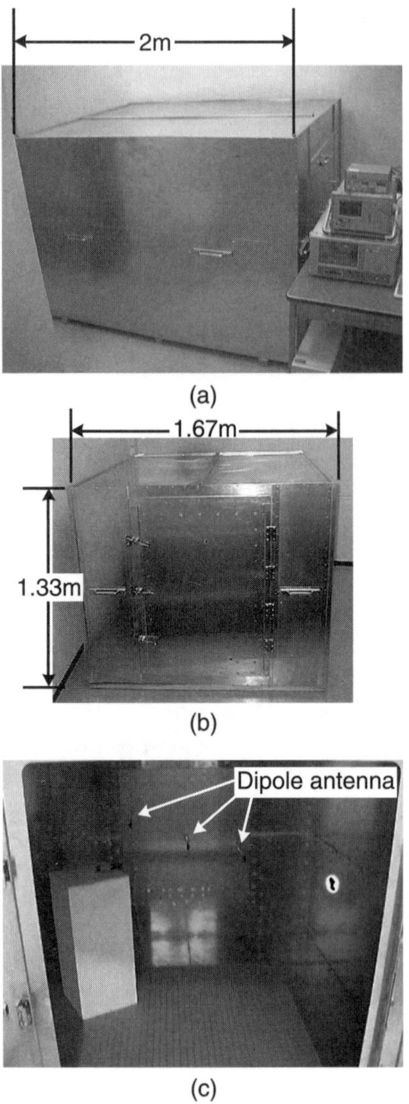

Figure 5.12 Photographs of field simulator: (a) side view, (b) front view, and (c) inside view.

wall opposite the source-mounted board, and its reflection coefficient is a function of the incident angle, the polarization, and the material. The reflection coefficient of a single-layer wave absorber can be derived by denoting the incident angle by θ, complex relative permeability and permittivity by ε_r and μ_r, respectively, and the absorber thickness by D as:

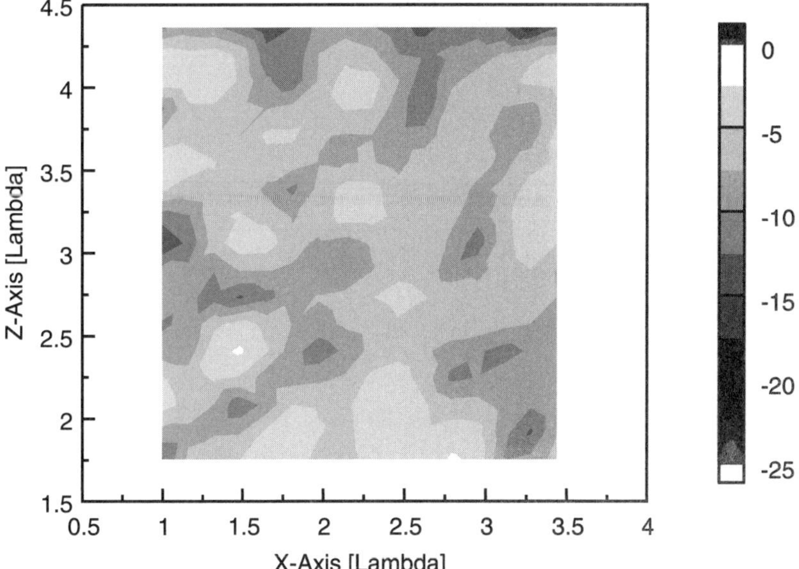

Figure 5.13 Calculated electric field distribution inside simulator.

$$S_{\mathrm{TE}} = \frac{Z_{\mathrm{TE}} - 1/\cos\theta}{Z_{\mathrm{TE}} + 1/\cos\theta}, \qquad S_{\mathrm{TM}} = \frac{Z_{\mathrm{TM}} - 1/\cos\theta}{Z_{\mathrm{TM}} + 1/\cos\theta} \tag{5.12}$$

$$Z_{\mathrm{TE}} = \frac{\mu_r}{\sqrt{\varepsilon_r \mu_r - \sin^2\theta}} \tanh\left(j\frac{2\pi D}{\lambda}\sqrt{\varepsilon_r \mu_r - \sin^2\theta} \right) \tag{5.13}$$

$$Z_{\mathrm{TM}} = \frac{\sqrt{\varepsilon_r \mu_r - \sin^2\theta}}{\varepsilon_{\mathrm{r}}} \tanh\left(j\frac{2\pi D}{\lambda}\sqrt{\varepsilon_r \mu_r - \sin^2\theta} \right) \tag{5.14}$$

where the electric field of the TM wave is on the incident plane and that of the TE wave is orthogonal to the incident plane. The frequency in the calculation is 820 MHz. Example electrical parameters for the wave absorber and the reflection characteristics are shown in Table 5.1 and in Figure 5.14, respectively. To calculate the reflection coefficients for both the TM and the TE waves, the incident wave is decomposed into each component. The incident plane is defined so as to include the transmitting point, the reception

Table 5.1
Example Electrical Parameters of Wave Absorber

Frequency	820 MHz
Permeability	$\varepsilon_r = 27.967 - 0.817$
Permittivity	$\mu_r = 5.171 - 3.590$
Thickness	6.3 mm

point, and the image source position, as shown in Figure 5.15, and n is the vector normal to the reflection plane.

$$\mathbf{E} = \frac{e^{-jk|\mathbf{r}|}}{|\mathbf{r}|}(S_{\mathrm{TM}}\ \mathbf{e}_{//} + S_{\mathrm{TE}}\ \mathbf{e}_{\perp}) \tag{5.15}$$

The electric field distribution calculated using (5.15) is shown in Figure 5.13 and its parameters are shown in Table 5.2. All transmitting sources are installed vertically, are excited with equal amplitude, and are all in phase. The contour map of the electric field strength near the wall (at a distance of less than λ) is excluded to avoid field divergence in the calculations. In the ray-tracing analysis, the reflection point is regarded as a secondary source, which makes $R \cong 0$ in the vicinity of the wall.

Before discussing fading characteristics, the cross-polarization ratio (XPR) inside the simulator is described. The results shown in Figure 5.16 were obtained by assuming that all the sources were mounted vertically. The inclined reflections from the wall, produced by the incident wave, slightly excite the horizontal polarization components. The XPR is 11.6 dB for Figure 5.16, and 12.3 dB for Figure 5.17 with all the sources mounted horizontally. This slight difference is caused by the difference between the simulator width and height. The variation in the range of XPR from a rotating transmitting source is about 12 dB, which is adequate because the measured XPR in cellular system is from 3 to 9 dB.

5.2.3.2 Cumulative Probability Distribution

To reproduce the fading propagation environment, the phases of all the transmitting antennas are varied randomly to vary the electric field distribution at the observation point. The fading distributions in the z-x plane at $y = 1.82\lambda$ are shown in Figure 5.18. Most of the points indicate Rayleigh or Rician distribution inside the field simulator. Each distribution is seen to spread out

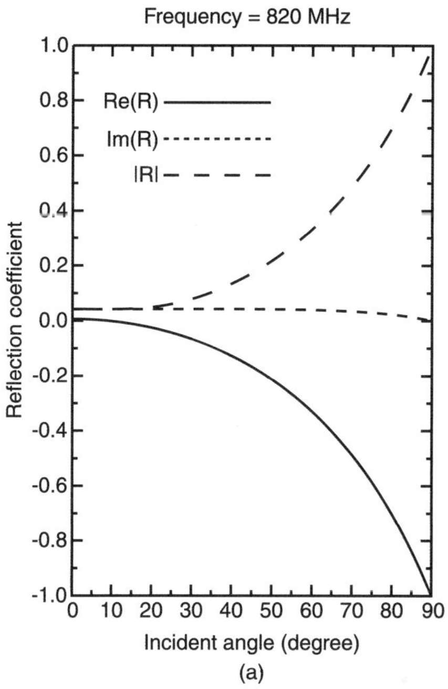

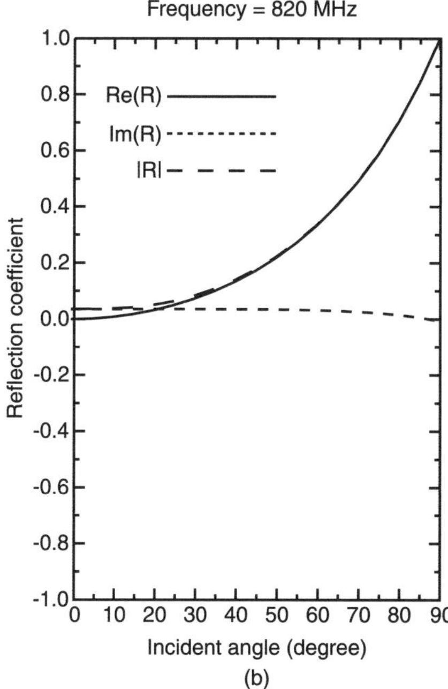

Figure 5.14 Reflection characteristics of wave absorber: (a) TE wave and (b) TM wave.

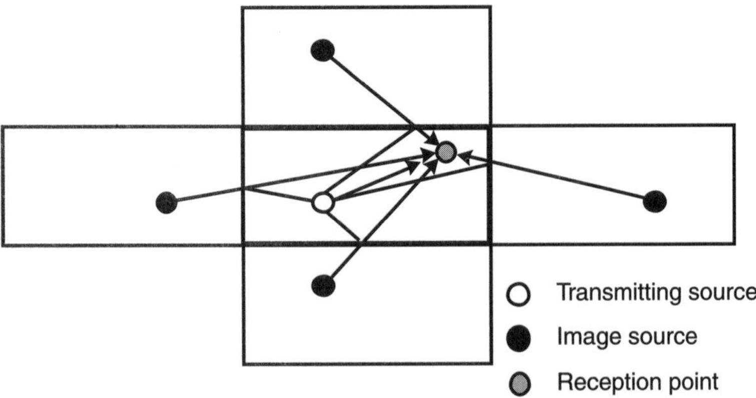

Figure 5.15 Reception point and image source position in *x*-*y* plane.

in the region of $0.5\lambda \times 0.5\lambda$. The Rayleigh distributions are observed on the source side for $z < 2.5\lambda$ and the rest of the area is dominated by Rician distribution. This is because reflection near the absorber is suppressed and the direct waves from the transmitting sources become the dominant components.

Figure 5.19 shows the cumulative probability distribution at the observation point for Rayleigh and Rician distributions with the electric field strength on the horizontal axis normalized by the median value. The distribution shown in Figure 5.19(a) follows the Rayleigh curve over the normalized electric field strength of –28 dB with a probability of 0.1%, which is

Table 5.2
Parameters for Calculation of Electric Field Distribution in Figure 5.13

Parameter	Value
Frequency	820 MHz
Simulator size (*x*,*y*,*z*)	4.51λ, 3.64λ, 5.47λ
Number of reflections *N**	5
Source 1 (*x*,*y*,*z*)	1.34λ, 2.18λ, 0.49λ
Source 2 (*x*,*y*,*z*)	2.26λ, 1.82λ, 0.49λ
Source 3 (*x*,*y*,*z*)	3.17λ, 1.64λ, 0.49λ
Observation plane	*z*-*x* plane at *y* = 1.82λ
Observation region in *x*	1.0λ to 3.5λ
Observation region in *y*	1.5λ to 4.5λ
Sampling interval	0.01λ

* *N* is a parameter in ray-tracing calculations.

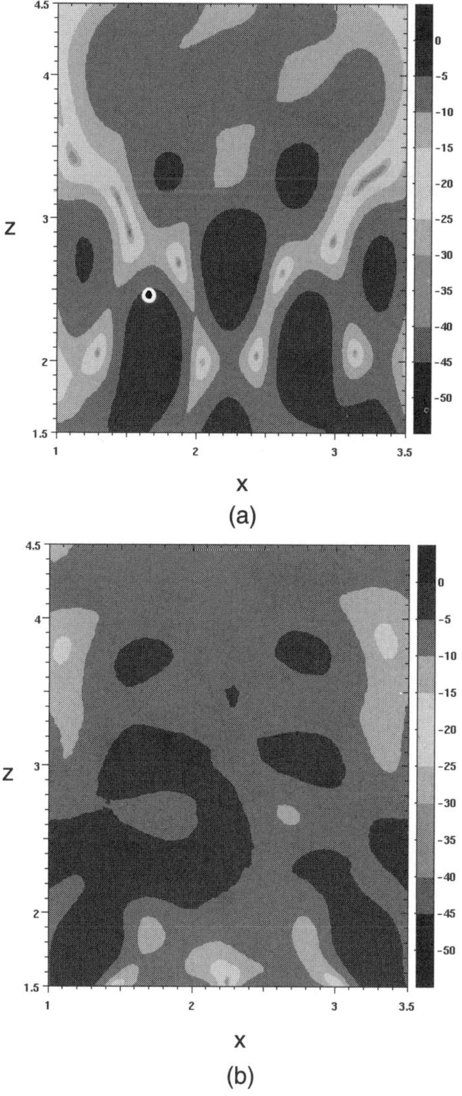

Figure 5.16 XPR inside simulator, with all sources mounted vertically: (a) vertical component and (b) horizontal component.

close to the result from outdoor measurements. The electric field strength as a function of time is shown in Figure 5.20(a). This time variation is the result of the change of phase at the source, and it is equivalent to the fading pitch. Figure 5.20(b) shows the results at half the switching speed of Figure 5.20(a). In these examples, both short-term medians are the same, with

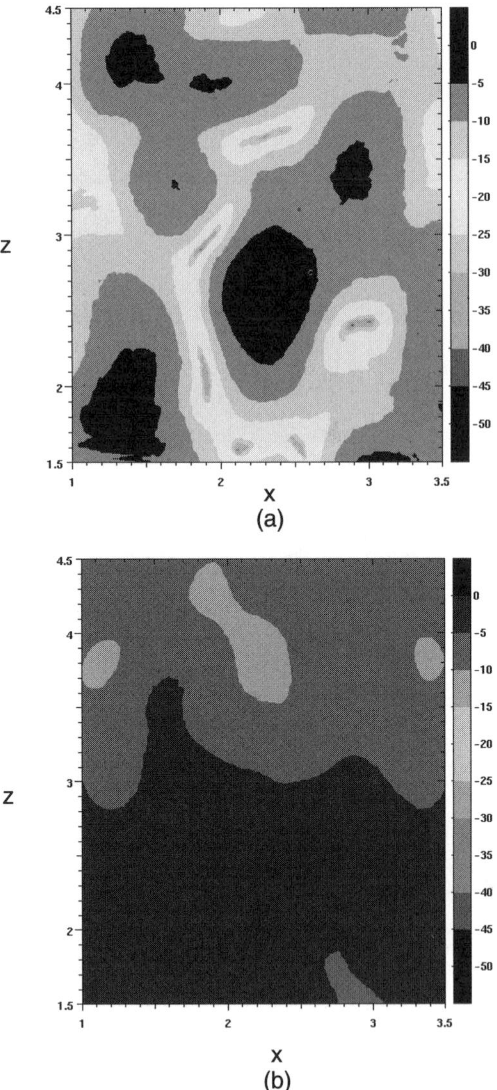

Figure 5.17 XPR inside simulator, with all sources mounted horizontally: (a) vertical polarization and (b) horizontal component.

values of –7.54 dB. Thus, the fading pitch inside the field simulator is controlled by the switching speed of the phase shifter.

The preceding discussion is based on the *z-x* plane as a two-dimensional case. Practical measurements on handset terminals require some volume to include a three-dimensional structure. The volume required to give a Rayleigh distribution is shown in Table 5.3, where the *z-x* position is shown in Figure

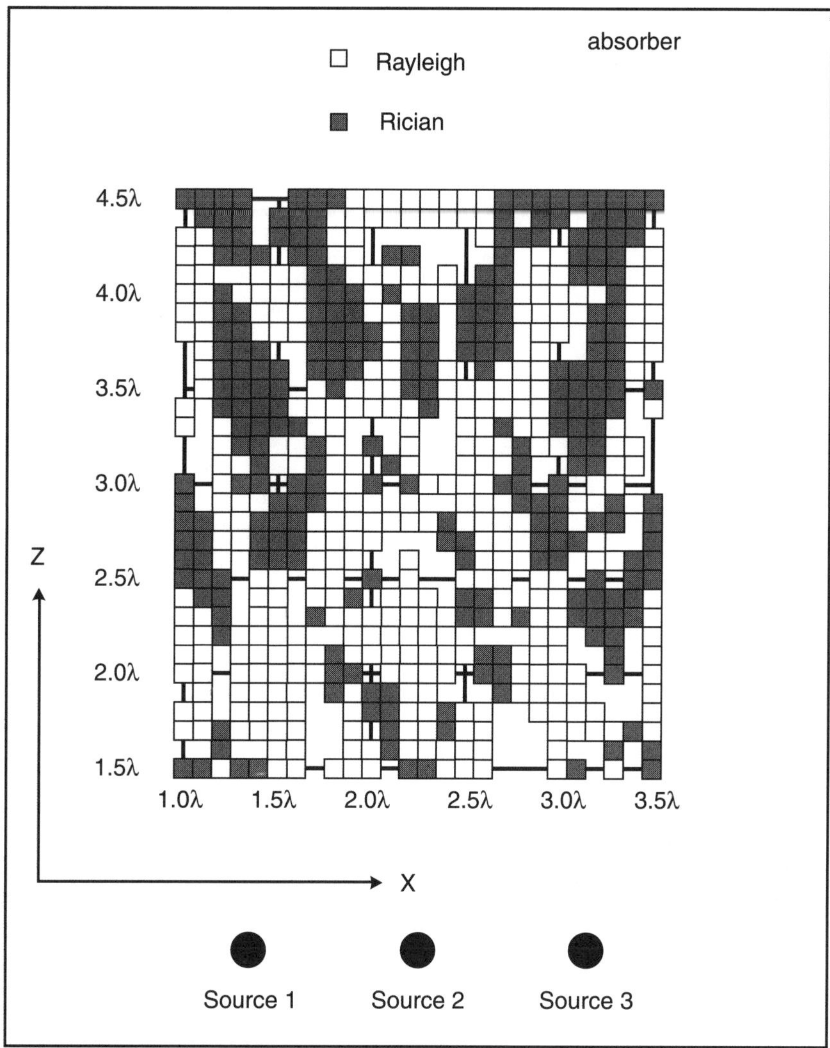

Figure 5.18 Fading distributions on the *z-x* plane at $y = 1.82\lambda$.

5.21. The volumes for A and B are 5,000 cm^3 and 10,000 cm^3, respectively, both of which are sufficient for handset terminal measurements. The range of variation in electric field strength inside volume B is –10 to 30 dB.

5.2.3.3 Correlation Coefficient for Diversity Measurements

An important parameter for diversity measurement using field simulation is the correlation coefficient between two observation points. The correlation coefficient R between two measured data is calculated as:

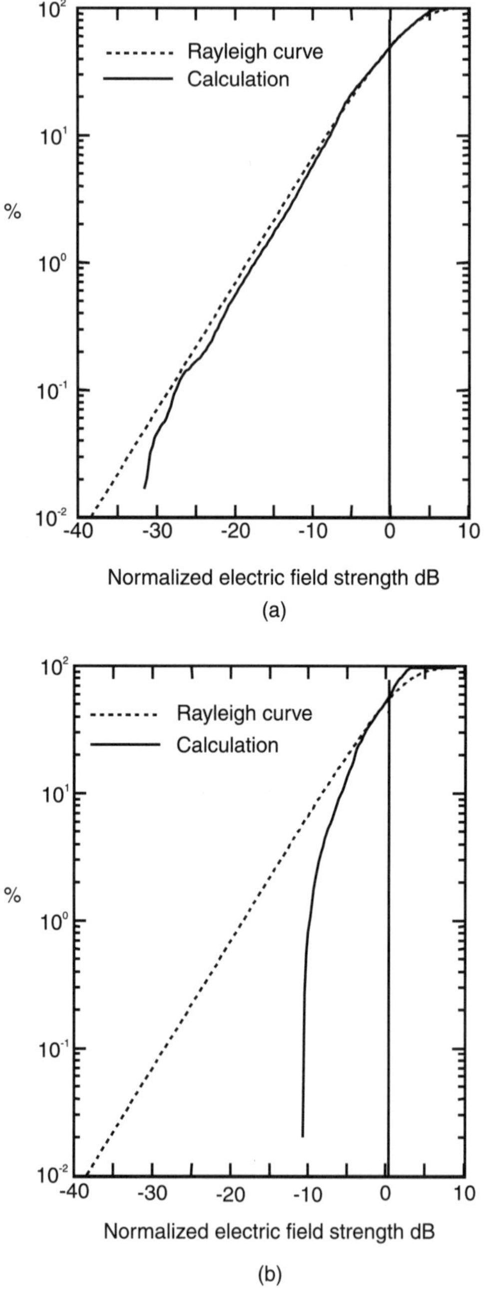

Figure 5.19 Cumulative probability distribution: (a) Rayleigh position and (b) Rician position.

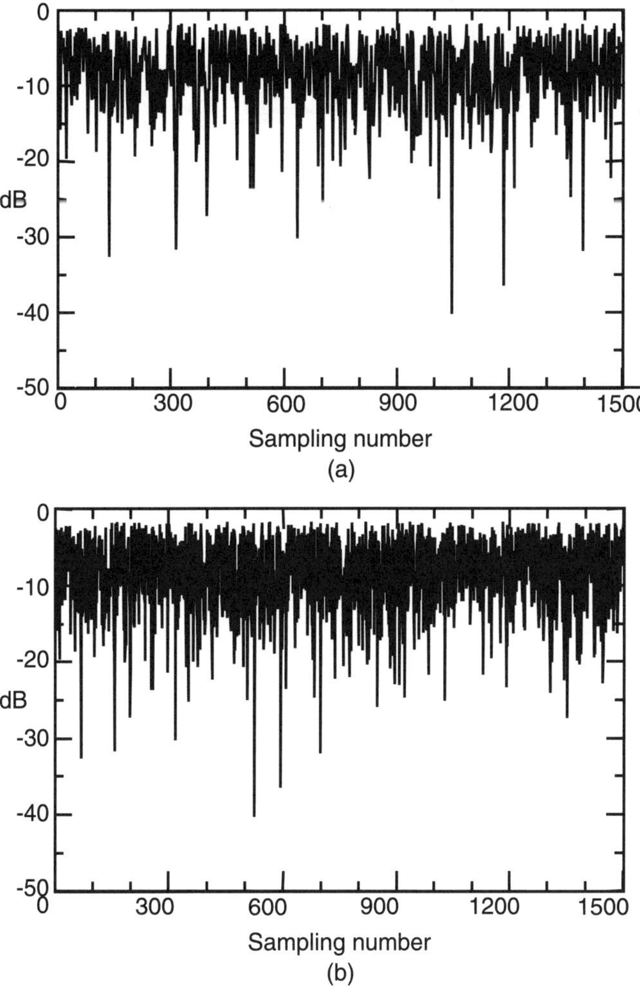

Figure 5.20 Electric field strength as a function of time: (a) $x = 2.3$, $z = 2.1\lambda$, and (b) $x = 2.3$, $z = 2.1\lambda$ where sampling pitch is half of that of part (a).

Table 5.3
Volume Size for a Rayleigh Distribution

	Volume (m^3) $x \times y \times z$
A	$0.35 \times 0.72 \times 0.40 = 0.10$
B	$0.30 \times 1.28 \times 0.35 = 0.13$
C	$0.25 \times 0.54 \times 0.20 = 0.03$

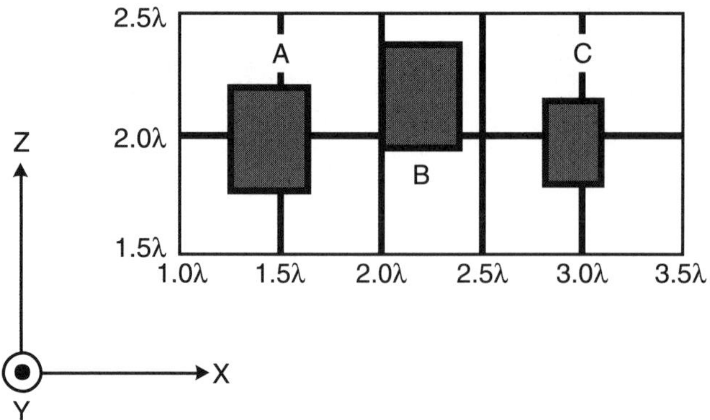

Figure 5.21 Area in *z*-*x* plane with Rayleigh distribution.

$$R = \frac{1}{(N-1)S_X S_Y} \sum_{I=1}^{N} (X_I - \bar{X})(Y_I - \bar{Y}) \tag{5.16}$$

where N is the number of data, $\bar{X}$, $\bar{Y}$ are average values of X, Y, and S_X, S_Y are the standard deviations, respectively. The range for the correlation coefficient R is $-1 \leq R \leq +1$. $R = 1$ means that X is proportional to Y, and $R = -1$ means that X is inversely proportional to Y.

Figure 5.22 shows the histogram of correlation coefficients in volumes A and B. This histogram is similar to that obtained from outdoor measurements, and its average value, between 0.4 and 0.2, is typical for real measured data.

5.2.3.4 Fundamental Measurements Using a Prototype Field Simulator

This section presents measured field distributions obtained from the prototype field simulator whose electrical performance was discussed earlier. The simulator consists of an aluminum wall and three standard dipole antennas as the transmitting sources.

To measure the electric field distribution inside the simulator, an electro-optical (E/O) probe measures the signal, and an optical fiber is used to transmit the signal after E/O conversion so as not to disturb the field distribution inside the simulator. In this measurement, three analog phase shifters are employed to change the phases of the signals from the transmitting dipole antennas. For simplicity, a digital phase shifter is not used in the system configuration. Figure 5.23(a) shows the electric field strength distribution with no phase variation. A standing wave is observed along the x and z axes. In

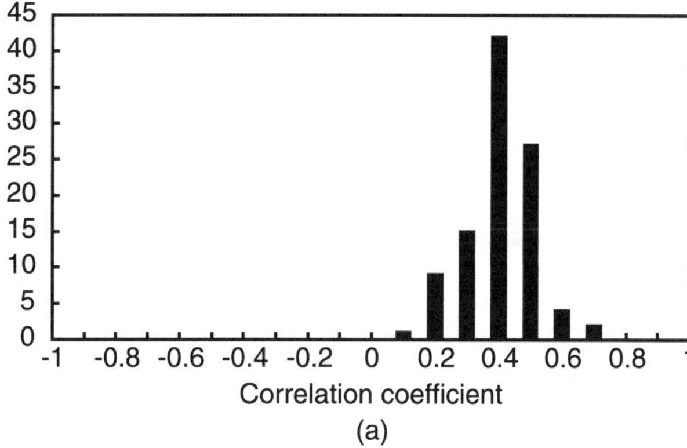

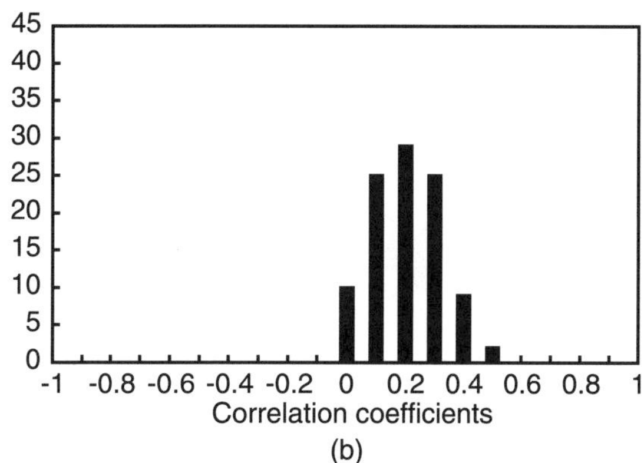

Figure 5.22 Histogram of correlation coefficients in volumes A and B: (a) area A, average value is 0.4; and (b) area B, average value is 0.2.

this example, the measurement probe is moved in 10-cm steps. Figure 5.23(b) shows the field distribution when the phase of the transmitting dipole is changed randomly, that is, it shows the case for a random field distribution inside the simulator.

The measured fading distribution is shown in Figure 5.24, from which it can be seen that the Rayleigh fading areas are approximately half as large as the Rician fading areas. The Rayleigh area is wide enough for useful phone measurements. In this example, the observation position should be moved to change the fading structure. It is convenient to make measurements with the handset under test placed at a fixed position for various fading structures.

The imbalances in the amplitude ratios for the transmitting sources allow the fading structure to be controlled, as shown in Figure 5.25. The

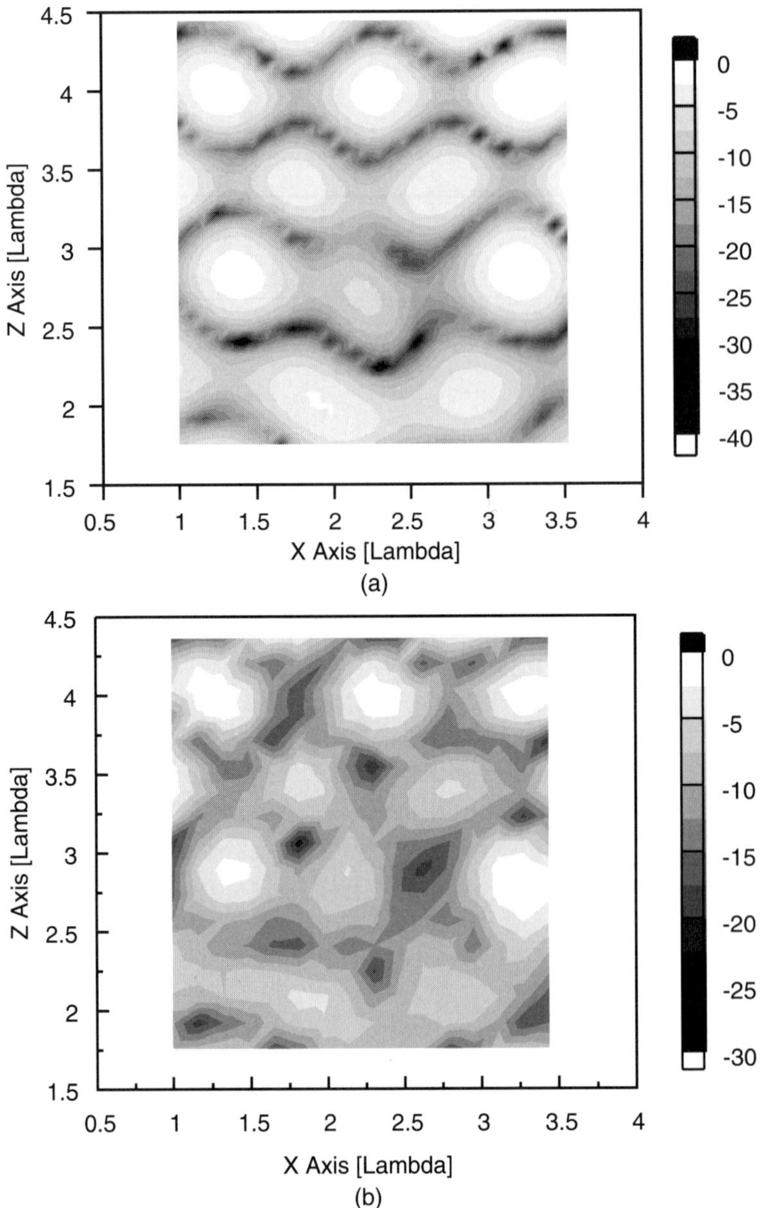

Figure 5.23 Measured electric field strength distributions (a) with no phase variation and (b) with phase changed randomly.

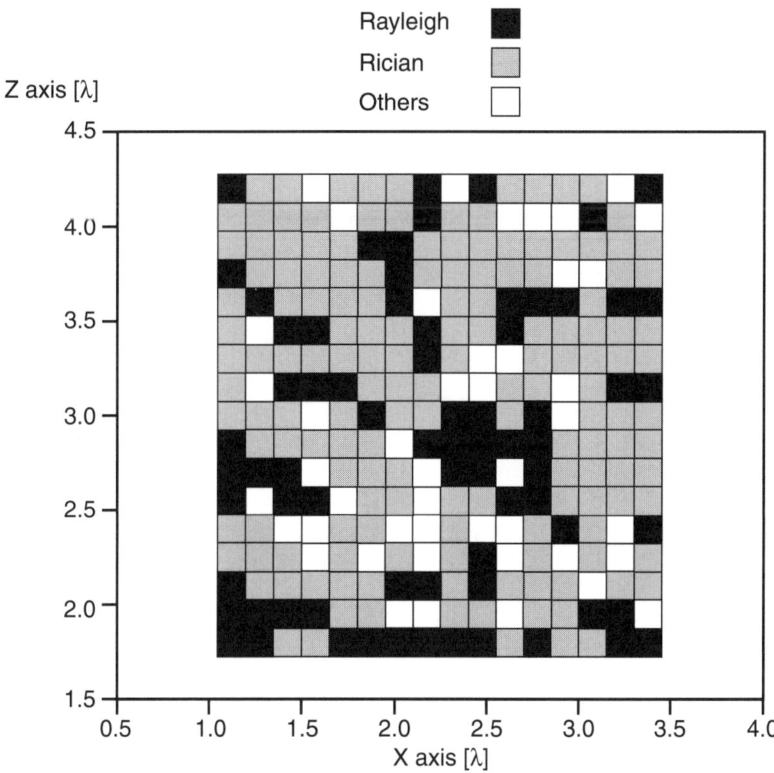

Figure 5.24 Fading distribution inside simulator.

amplitude ratios for the sources, which vary from 1:1:1 to 1:4:1 change the fading structure at the fixed observation point from Rayleigh to Rician. Thus, amplitude control is effective in changing the fading structure inside the field simulator.

5.3 Sensitivity Measurements Using a Field Simulator

Because the movement of portable terminals subjects them to severe fading in a multipath propagation environment, especially in the city, it is insufficient for tests on handsets to evaluate only static characteristics in a typical anechoic chamber, as described in Chapter 2. Here, the term *static characteristics* refers to the measurement of the input impedance and the radiation pattern of handset antennas. The random field method, which actually involves measurements on handsets in an outdoor propagation environment,

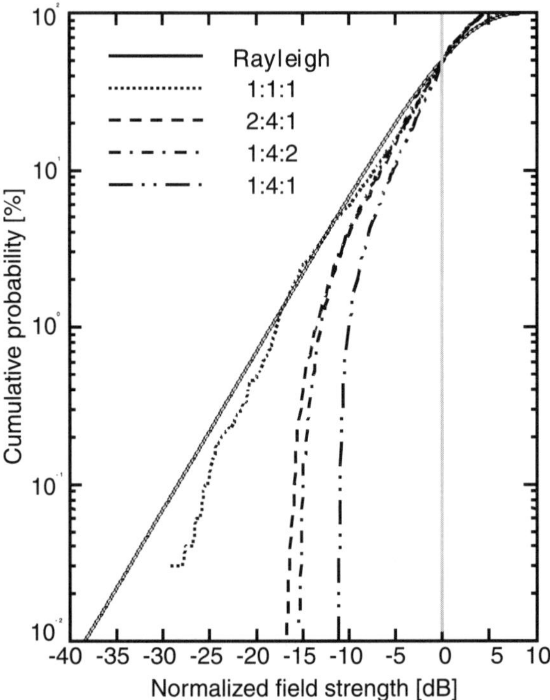

Figure 5.25 Fading structure change due to imbalance in amplitudes of sources.

is time-consuming on a large scale, gives poor repeatability of measured results, and requires an appropriate radio license. In the daytime, the measured electric fields are affected by human activity and the surrounding traffic conditions. The field simulator described in the preceding section provides a solution to these problems since it is a portable handset evaluation system that reproduces the artificial fading environment inside a building.

The particular field simulator presented earlier is slightly smaller and is only used to evaluate handset terminals. In practice, handset terminals are always held in the hand and are used close to the human body. The performance of handset terminals should therefore be tested in conjunction with an actual human body or a phantom that provides electrically equivalent performance. In addition, the final test of handset products requires their manipulation by human carriers to confirm that the functions really work. To evaluate a handset terminal with a human body inside a field simulator, it should be larger than the model discussed in Section 5.2. The size of the field simulator is enlarged to enclose the whole human carrier for the handset test in the example in this section.

Due to intense competition in the cellular phone market, each manufacturer is being asked to develop miniaturized handsets with high sensitivity in a short time frame. Close attention must therefore be paid to the evaluation of handset characteristics, including the antenna, and the use of field simulators has been presented as a means of solving these problems. Because the characteristics of an outdoor multipass environment can be reproduced indoors, the quantitative evaluation of the characteristics of handset terminals can be carried out in a field simulator. An important factor in the evaluation of a handset is its effective sensitivity to receive weak RF signals. The sensitivity is measured directly by inputting the RF signal from a signal generator to the antenna input port. This measurement requires removal of the casing from the handset antenna, but this procedure is not carried out during the final evaluation test of the product. If attention is not paid to the multipath fading effect during the evaluation stage, the sensitivity can be measured inside an anechoic chamber. In general, very weak reception levels are often observed around a cell edge and in the shadows of buildings. Because these propagation environments are affected by multipath fading, the sensitivity for the handset can be evaluated inside a field simulator.

The sensitivity of handset terminals is verified not only by the minimum usable RF input signal level, but also by using controlling software to determine the timing of reception in the digital cellular system. To confirm the effectiveness of this software, the number of successful connections is the best evaluation parameter for the sensitivity performance of handsets, as described in Section 2.6.1.3. A test under weak reception field strength conditions is also important for the handover function in current digital cellular systems. In the present system, the handset terminal assesses the switching timing from the current base station to the next as the handover function. This overlapping region between two cells is located around the cell edge and is in the weak field strength region.

A public digital cellular (PDC) system in Japan using time-division multiple-access (TDMA) with a narrow frequency band has an optional function of diversity reception for the handset terminals. A sleeve antenna and a built-in antenna give this diversity reception, and its diversity characteristics are verified by measuring the complex radiation pattern, as described in Section 2.4. This diversity measurement gives only the correlation coefficient between two antennas, and the diversity performance is evaluated using the theory described in Section 1.3.3, while assuming Rayleigh distribution for the propagation characteristics. As described in the preceding discussion on sensitivity, the handset performance is not determined by its hardware characteristics but by its control software. Therefore, the diversity reception should be evaluated for the handset terminal including all the hardware and software in a multipath propagation environment, such as a field simulator.

This section presents a field simulator that can be applied to sensitivity evaluation for handset terminals in a weak electric field environment. In addition, the diversity characteristics are also evaluated for the handset antennas.

5.3.1 Construction of a Field Simulator

The overall arrangement of the field simulator presented in this section is shown in Figure 5.26. The handset under evaluation is mounted on a board hung from the ceiling. The field simulator consists of a cube having sides of 2m for measurements with a frequency limit of 800 MHz. This size is adequate to accommodate a human body to manipulate the handset inside the simulator, as shown in Figure 5.27.

All outer walls surrounding the measurement field consist of metal, and three standard dipole antennas are arrayed, offset from the source. An absorber-covered wall is not used in this example because the size of the simulator is large. A block diagram of the field simulator is shown in Figure 5.28. The transmitting signal is distributed from the source to the three antennas with an arbitrary phase relationship controlled by a personal computer (PC), and produces a fading environment by disturbing the electric field inside the measurement area.

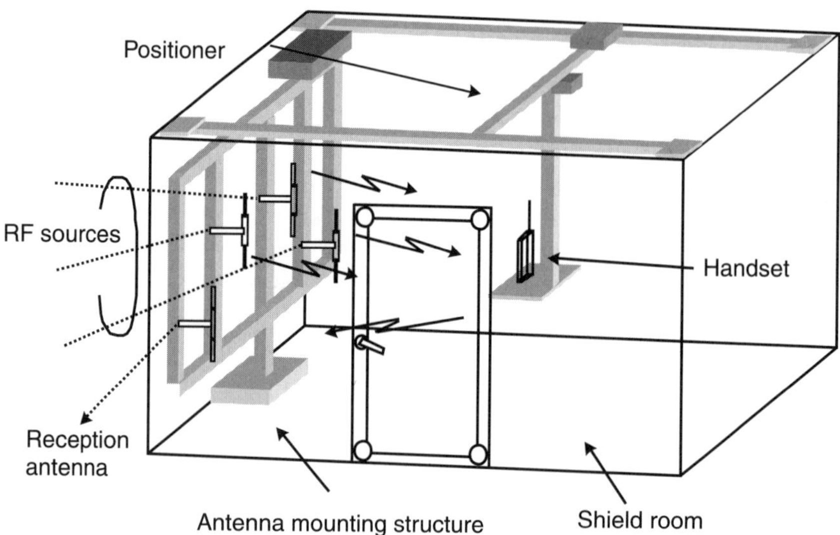

Figure 5.26 Outline of the field simulator.

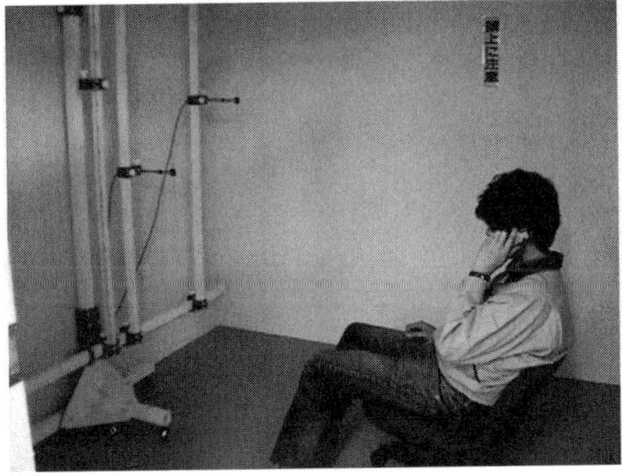

Figure 5.27 Photograph of interior of field simulator.

Fading generators are used to change the phases at random and produce multipass fading such as Rayleigh and Rician fading independently. Two fading generators are used to examine the handover characteristics of the handset terminal by receiving the electric fields with the directional antenna turned toward two different base stations. In general, the arrangement uses the base station signal for the RF input, but, in this case, a base station simulator is used instead of the actual base station.

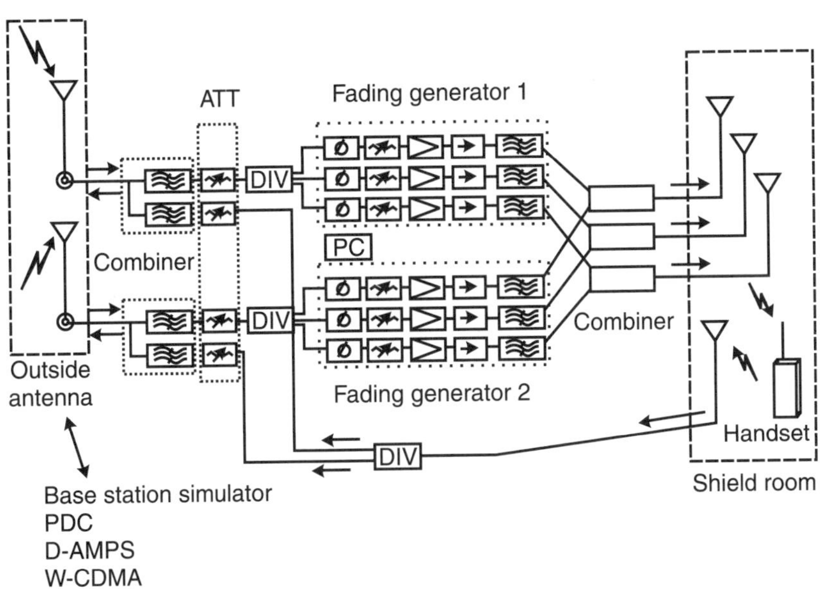

Figure 5.28 Block diagram of field simulator.

The fading generator consists of an eight-bit digital phase shifter to change the antenna phase as a function of time. This simulator can also be used to examine the handset for TDMA systems with two RF signals having different frequencies. The two RF signals are combined and input to the antennas inside the simulator. The variable attenuator inserted in front of the fading generator adjusts the level of the transmitting RF signals and allows simulation of arbitrary weak and medium strength electric fields, which correspond to the conditions in an outside propagation environment.

Although an actual base station is used in this example, a base station simulator is substituted for the open air antenna part when a new mobile communication system is introduced that is not intended to be used in the country where the field simulator is installed. In this case, the field simulator is considered to be merely an attenuator in the propagation path and, therefore, the system is free from rules and regulations.

The received electric field strength at the handset under test is picked up by a reception dipole antenna installed inside the simulator, as shown in Figure 5.28. The reporting signals from the handset are RSSI level and the number of successful connections.

5.3.2 Test for the Number of Successful Connections

Figure 5.29 shows the cumulative probability distributions measured both inside the field simulator and outside in the weak electric field strength areas. Two examples of outside areas are the suburbs and the countryside, where the measured electric field strengths are in the range of 5 to 15 dBμ. The suburban area is located at the cell edge in a medium size city, and the countryside is at the cell edge of a lakeside. Both measured probability distributions inside the simulator became nearly equal to those for outdoor conditions, which shows the effectiveness of the field simulator from the viewpoint of fading distribution.

The measured number of successful connections in both the field simulator and the outdoor area are summarized as a percentage success rate in Table 5.4. Three commercial handsets have been evaluated in these measurements. The total number of connections was 50 for the outdoor measurements and 100 in the field simulator. The number of measurements in the outdoor area was limited by the measurement time required. The handset antenna was used in the extended position, and the number of successful connections was measured in the paging mode. The difference in the success rates for measurements in the outdoor area and in the simulator is less than 5%, which clearly shows that measurements using the simulator can replace outdoor

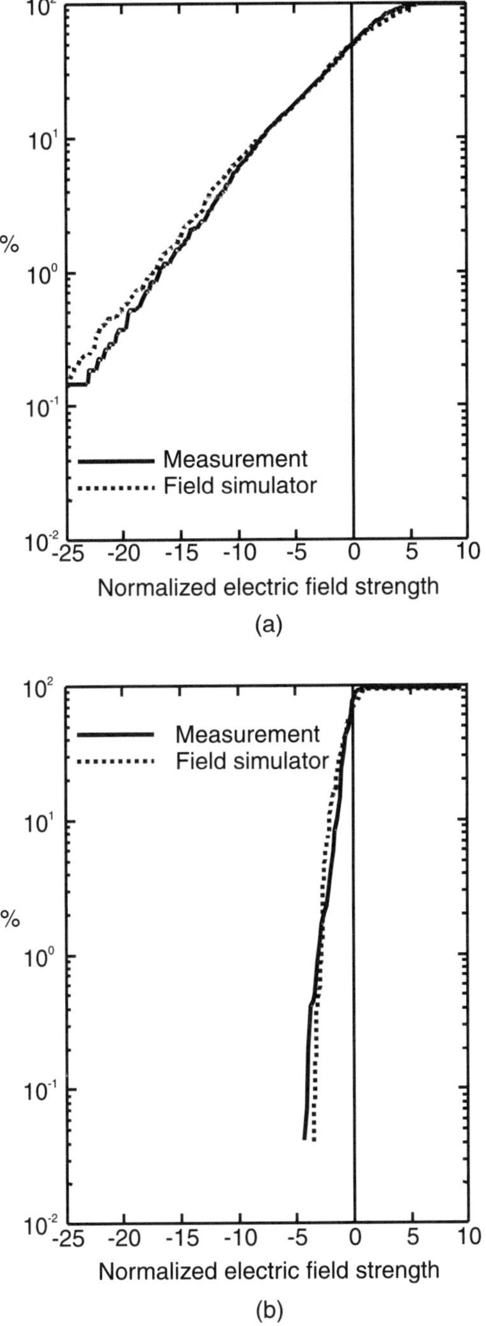

Figure 5.29 Cumulative probability distributions measured inside the field simulator: (a) data from the suburb simulation and (b) data from the countryside simulation.

Table 5.4
Successful Connection Rate Percentage in Actual Field Test and Inside the Field Simulator

Successful Rate	Handset A	Handset B	Handset C
Shikotsu Lake	92%	80%	94%
Field simulator	90%	85%	94%

measurements in this connection test. The measured success rates indicate the difference in sensitivities of the handsets.

5.3.3 Diversity Performance Measurements

This section presents diversity performance measurement for handsets with a whip antenna and a built-in antenna, using the field simulator. A handset and its antennas are shown in Figure 5.30. The whip antenna is a half-wavelength sleeve antenna, and the built-in one is a planar inverted-F antenna (PIFA) with a notch for impedance matching. This combination of

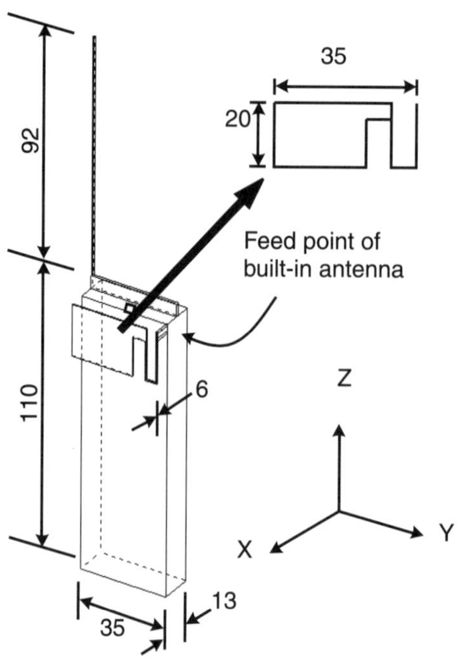

Figure 5.30 Handset and its antennas used for measurement: all the units are in millimeters.

whip antenna and PIFA is widely used for reception diversity in handset terminals [13]. The measured radiation patterns in the vertical plane are shown in Figure 5.31. The differences in the radiation patterns result in low values for the correlation coefficients for diversity reception.

The diversity performance is evaluated by measuring the cumulative probability distribution inside the field simulator. The definition of diversity gain is given in Section 1.3.3. The diversity performance of digital cellular terminals is evaluated by how much the bit error ratio (BER) is improved, as described in Section 5.1.4. This measurement is applicable also to the field simulator by reporting the BER from the handset. In this section, an example of diversity gain is presented.

The measured data as a function of time obtained with the above handset inside the field simulator is shown in Figure 5.32, for a total of 10,000 data points for each antenna. The median values of the sampled data for both the whip and built-in antennas are normalized at 0 dB. Several instances of deep fading are apparent for both antennas. The RSSI level for each antenna is reported simultaneously as the uplink signals in this measurement. As shown in Figure 5.32, the time during which deep fading occurs for the whip antenna is different from that for the built-in antenna. This is because of the diversity effect and is confirmed quantitatively by evaluating the diversity gain of this handset antenna arrangement.

Figure 5.33 shows the cumulative probability distribution normalized by the median value of the built-in antenna. The fading structure for each antenna is very close to the Rayleigh distribution, and the diversity gain is about 7 dB with a cumulative probability of 1%. The diversity gain is based on selection diversity. In selection diversity, the antenna with the highest power combines the data sequences that represent the receiving electric field strength from each antenna. The correlation coefficient in this example, calculated using (5.16), is 0.44.

With this antenna arrangement, the diversity gain can be increased by changing the position of the built-in antenna. When the built-in antenna is at the center of the casing, the measured diversity gain is 11 dB and the correlation coefficient is –0.35. Although this antenna location is not realistic because the built-in antenna is covered by the hand when in use, the diversity antenna can be evaluated quantitatively from field simulator measurements.

5.4 Delay Spread Measurement Using a Field Simulator

Mobile communication services are used not only for voice signals, but also for high-speed digital data transmissions, because the mobile network can be

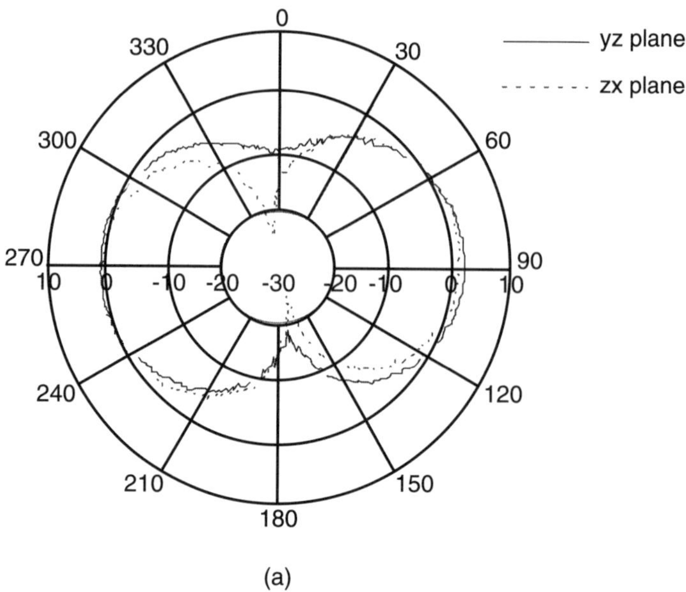

(a)

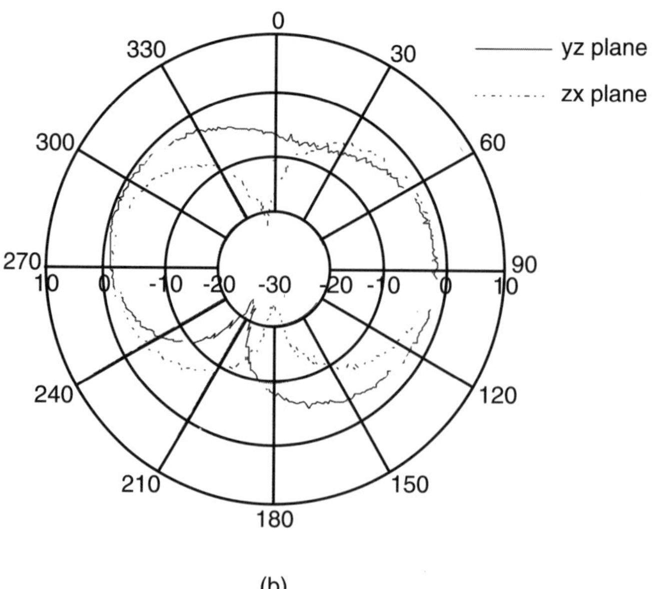

(b)

Figure 5.31 Measured radiation patterns in vertical plane of handset: (a) whip antenna and (b) built-in antenna.

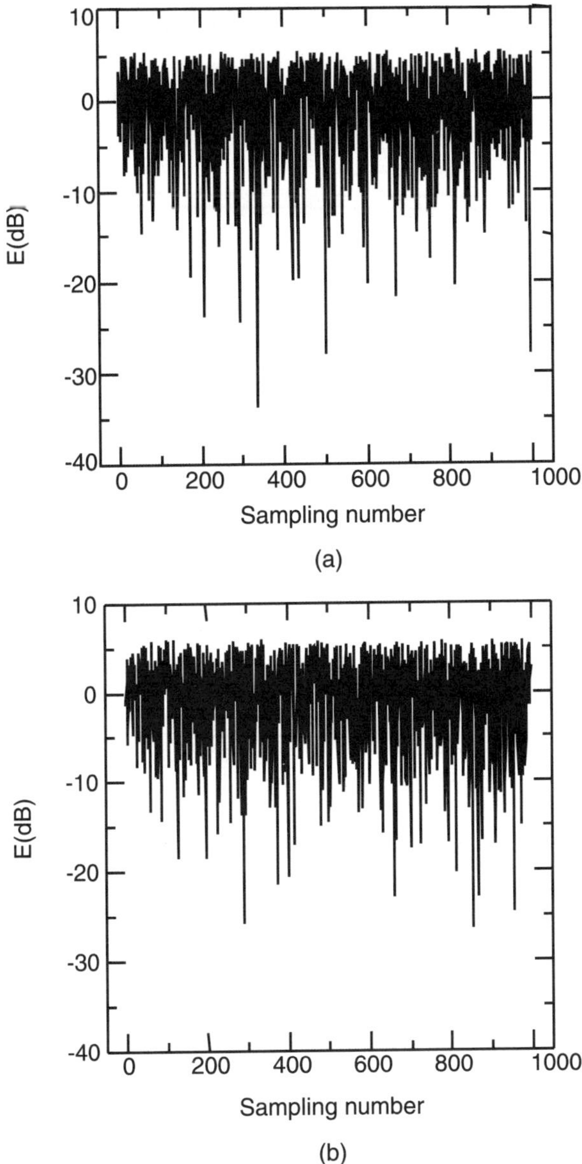

Figure 5.32 Measured data as a function of time with handset inside the field simulator: (a) whip antenna and (b) built-in antenna.

connected to the Internet. In the code-division multiple-access (CDMA) system, electromagnetic waves arriving at a reception point with time delays are synthesized to increase the level of reception signals, using a Rake receiver. It is necessary, therefore, for the performance of mobile terminals used in this

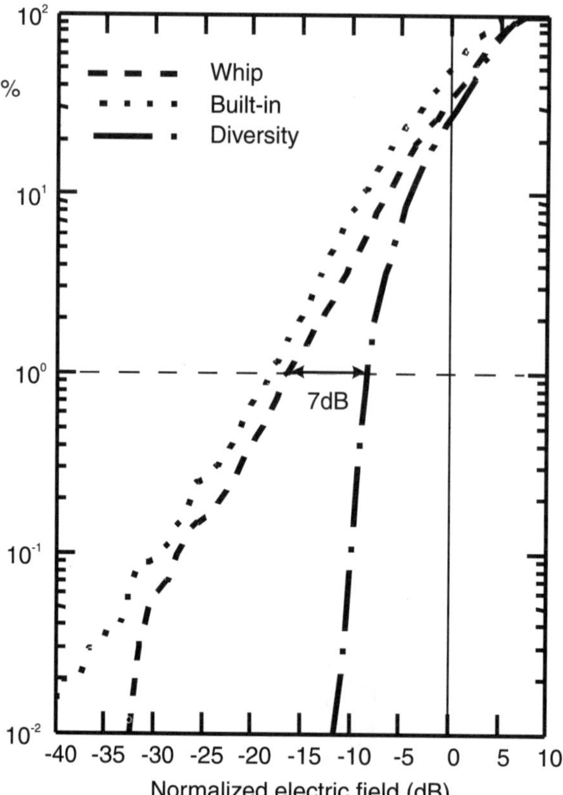

Figure 5.33 Cumulative probability distribution.

system to be evaluated by inputting several RF signals with different time delays. Signal processing for signals delayed in time is carried out at baseband frequency and can be examined using the fading simulator described in Section 5.1.

As the number of mobile terminals and personal digital assistants (PDAs) increases as a proportion of total digital radio data transmission systems, they are required to be evaluated as products before shipping with their antennas. A field simulator with a delay spread function meets this requirement, but this is still under development. This section first presents theoretical models for delay spread based on propagation measurement data, and then describes a possible method for delay spread measurement in a field simulator.

5.4.1 Delay Profile Model

To evaluate the characteristics of the delayed waves defined in Section 1.4.1, typical models for the delay profiles, given by simple equations, are required [14]. Figure 5.34 shows four typical models for delay profiles. A double spike profile, as shown in Figure 5.34(a), is a simple model that assumes two waves with an average power level P_i ($i = 1, 2$). This model is widely used for the analysis of data transmissions using a fading simulator. A delay profile as a function of time τ can be obtained using the delta function $\delta(x)$ as:

$$p(\tau) = \frac{1}{P_1 + P_2}\left\{P_1\delta(\tau - \tau_1) + P_2\delta(\tau - \tau_2)\right\} \tag{5.17}$$

The power level P_i and the time interval of the two waves, $\Delta\tau = \tau_2 - \tau_1$, are defined so that the BER in the simulation becomes the same as the data from propagation measurements.

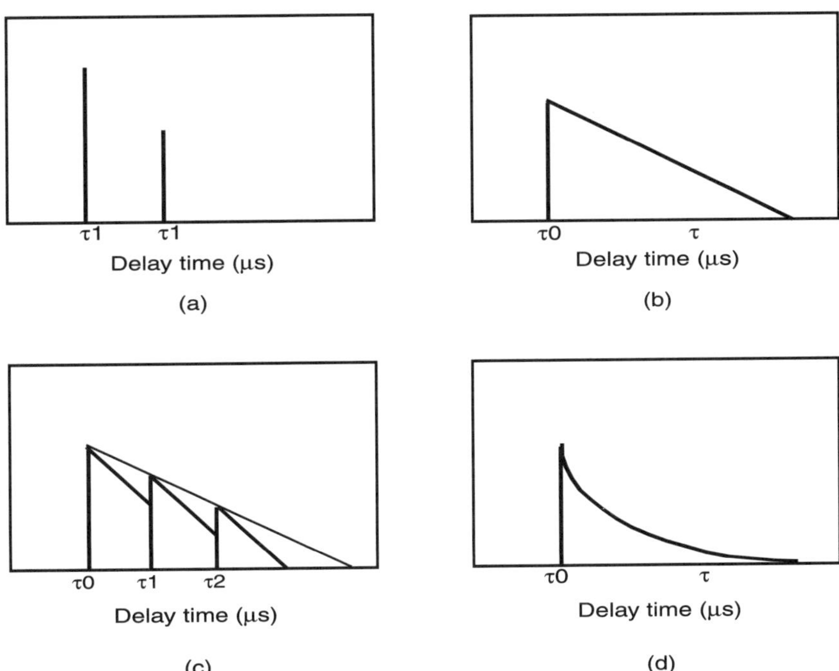

Figure 5.34 Typical models for delay profiles: (a) double-spike model, (b) exponential function model, (c) multiple exponential function model, and (d) power series function model.

A model for simulation of the delay profile in cities can be derived using the exponential function. Its profile can be expressed as (5.18) by denoting the delay spread by σ_o, and it is normalized to the power of 1, and integrated in the time domain as:

$$p(\tau) = \frac{1}{\sigma_o} \exp\left(-\frac{\tau - \tau_o}{\sigma_o} \right) U(\tau - \tau_o) \tag{5.18}$$

where $U(x)$ is a unit function and $U(x) = 1$ if $x \geq 0$, and 0 if $x < 0$. The characteristics calculated using this profile can be made nearly equal to the measurement data by choosing an appropriate value of σ_o between 1 and 2 ms.

The above two profiles are expanded in the following equation as a model using multiple exponential functions:

$$p(\tau) = \frac{1}{\sum_{i=0}^{n-1} \sigma_i} \sum_{i=0}^{n-1} \exp\left(-\frac{\tau - \tau_i}{\sigma_i} \right) U(\tau - \tau_i) \tag{5.19}$$

This profile is a summation of (5.18) and is equivalent to (5.18) when $n = 1$. It is also equivalent to the double-spike model of (5.17) when $n = 2$ and $\sigma_i = 0$. This profile can include several delayed large waves, as often occurs in a basin surrounded by high mountains and at some locations surrounded by tall buildings in a big city.

In addition to the profile models just given, the following equation expresses propagation characteristics for a microcellular system with a coverage of 100 to 500m. The base station in this system is assumed to be as high as surrounding buildings. The profile function is given as:

$$p(\tau) = \frac{\alpha - 1}{\tau_o^{-\alpha+1}} \tau^{-\alpha} U(\tau - \tau_o) \tag{5.20}$$

where the constant α is 3 or 4.

The average power of delayed waves and the delay spread can be obtained analytically from (5.17) to (5.20) and are summarized in Table 5.5. Those of (5.19) have been calculated numerically.

Table 5.5
Average Power of Delayed Waves and Delay Spread

Model	Average Power	Delay Spread
(a)	$\dfrac{P_2(\tau_2 - \tau_1)}{P_1 + P_2}$	$\dfrac{\sqrt{P_1P_2}}{P_1 + P_2}(\tau_2 - \tau_1)$
(b)	σ_0	σ_0
(c)	$\dfrac{1}{\alpha - 2}\tau_0$	$\tau_0\sqrt{\left(\dfrac{\alpha - 1}{\alpha - 3}\right) - \left(\dfrac{\alpha - 1}{\alpha - 2}\right)^2}$

5.4.2 Reproduction of the Delay Profile Using a Field Simulator

It is necessary to reproduce the propagation characteristics with delayed waves described in the above, but the interval of delayed waves in the field simulator is not equal to that for actual measured data due to limitations in the size of the simulator. A field simulator with a maximum size of 1 to 2m gives delayed waves of the order of 10 ns. On the other hand, outdoor delayed waves are of the order of 10 μs, which indicates a major difference with those of the simulator. In theoretical calculations assuming a field simulator with perfect conducting walls, electromagnetic waves inside are never attenuated and do not result in a delay profile whose amplitude decreases with time. In a real field simulator, however, a small degree of attenuation is observed at the walls due to the use of imperfect conducting material and small gaps at the connection points of the simulator. The above attenuation cannot be controlled, and is therefore not appropriate when measurements are made to obtain the delay profile described in the preceding section.

Figure 5.35 shows the delay profile inside a simulator, obtained using the ray-tracing method in a simulator having the parameters given in Table 5.6. In this example, a wave absorber covers all walls inside the simulator, giving a 25-dB absorption for normally incident plane waves. This profile applies to a case with a direct wave as well as a delayed wave caused by reflections from the side walls, and is similar to the double-spike model. Unfortunately, the time interval of the delayed waves is determined by the size of the simulator and cannot be controlled at will.

As a further example, suppose that the simulator with side walls covered by an absorber, as shown in Figure 5.36, has a delayed profile for which the amplitude decreases with time. This decreasing profile is close to the

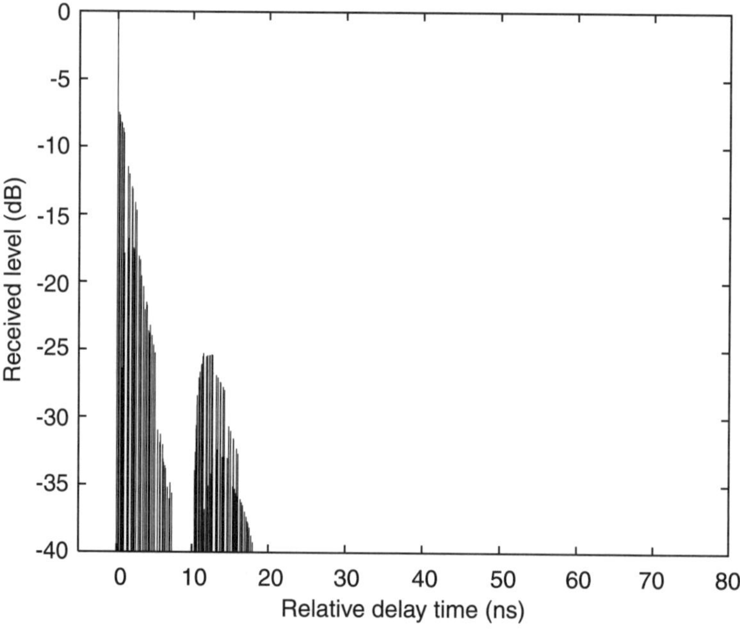

Figure 5.35 Calculated delay profile inside a simulator in which all walls are covered with a wave absorber.

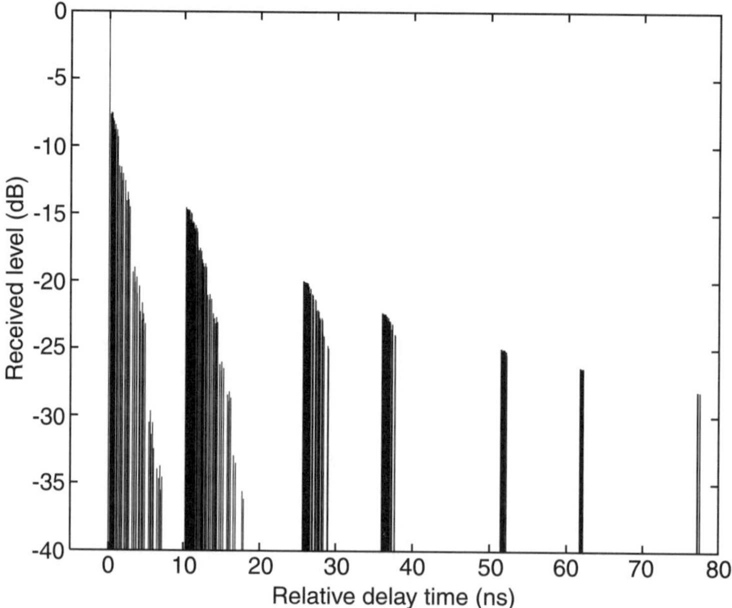

Figure 5.36 Calculated delay profile inside simulator of side walls covered with absorber.

Table 5.6
Parameters for Calculation of Delay Profile

Parameter	Value
Simulator size (x,y,z)	10.3λ, 10.3λ, 10.3λ
Number of reflections N*	7
Source 1 (x,y,z)	1.54λ, 5.15λ, 2.58λ
Source 2 (x,y,z)	2.04λ, 5.15λ, 2.58λ
Source 3 (x,y,z)	4.64λ, 5.15λ, 4.12λ
Observation plane	z-x plane at $y = 1.82\lambda$

* N is a parameter in ray-tracing calculations.

exponential function and its component in the horizontal axis is of the order of nanoseconds. This profile can be regarded as a single spike, because the actual delay time is of the order of microseconds. Therefore, this field simulator configuration enables an arbitrary delay profile to be obtained by inputting RF signals with time delays into the source antennas.

This type of field simulator has been considered only theoretically so far, and its demonstration in prototype form is left to the future.

References

[1] Arrendondo, G. A., et al., "A Multipath Fading Simulator for Mobile Radio," *IEEE Trans. Communication,* Vol. COM-21, No. 11, Nov. 1973, pp. 1325–1328.

[2] Hirade, K., et al., "Fading Simulator for Land Mobile Radio Communication," *Trans. IEC Japan*, Vol. 58B, No. 9, Sept. 1975, pp. 449–456.

[3] Clark, R. H., "A Statistical Theory of Mobile-Radio Reception," *Bell Syst. Tech. J.*, Vol. 47, No. 6, Jul.–Aug., 1968, pp. 957–1000.

[4] Arnold, H. W., and W. F. Bodtmann, "A Hybrid Multichannel Hardware Simulator for Frequency-Selective Mobile Radio Paths," *IEEE Trans. Communication*, Vol. COM-31, No. 3, 1983, pp. 370–377.

[5] Ueda, T., and H. Suzuki, "Selective Fading Simulator for Digital Mobile Radio," *Proc. IEICE Fall Conf.*, Sept. 1998, B-1-257 (in Japanese).

[6] RCR Std-27A, Association of Radio Industries and Businesses, Japan.

[7] Arai, H., et al., "A Flat Energy Density Antenna System for Mobile Telephone," *IEEE Trans. Vehicular Technology*, Vol. 40, No. 2, May 1991, pp. 483–486.

[8] Kuwabara, M., *Vehicular Telephone System,* edited by IEICE, Japan, Corona Co., 1985, p. 17 (in Japanese).

[9] Arai, H., "Field Simulator for Reproduction of Propagation Environment," *Proc. 1995 USNC/URSI Radio Science Meeting Conf. Digest*, Newport Beach, CA, June 1995, p. 318.

[10] Arai, H., "Field Simulator for Rayleigh/Rician Fading Reproduction," *Proc. IEEE AP-S Int. Symp. Conf. Digest,* Baltimore, MD, July 1996, pp. 1552–1555.

[11] Yamada, Y., et al., "Diversity Antennas for Base and Mobile Stations in Land Mobile Communication Systems," *IEICE Trans.*, Vol. E74, No. 10, Oct. 1991, pp. 3202–3209.

[12] Ohnishi, N., et al., "Field Simulator for Testing Handset Under Multipath Propagation Environment," *Proc. IEEE AP-S Int. Symp. Conf. Digest,* Montreal, Canada, July 1997, pp. 2584–2587.

[13] Taga, T., and K. Tsunekawa, "Performance Analysis of a Built-In Planar Inverted F Antenna for 800 MHz Band Portable Radio Unit," *IEEE Trans. Selected Areas in Communications,* Vol. SAC-5, No. 5, June 1987, pp. 921–929.

[14] Fujii, T., "Mobile Propagation Characteristics of Wideband Transmission," *NTT DoCoMo Tech. J.*, Vol. 7, No. 4, Jan. 2000, pp. 60–70 (in Japanese).

Appendix A
Conversion of dBμV to dBm

The received electric field strength at the antenna input port is measured as a voltage V_μ in units of dBμV. Here we describe how it can be converted to units of received power P_r in dBm.

When the impedances of both the antenna and the receiver input ports are matched to Z_o, the impedance of the transmission line connecting the antenna and the receiver P_r is expressed by V (V) as:

$$P_r \text{ (dBm)} = 10 \log_{10}\left(\frac{V^2}{4Z_o} \times \frac{1}{10^{-3}}\right) = 20 \log V - 10 \log Z_o + 30 \quad \text{(A.1)}$$

The voltage is given as a unit of dBmV as:

$$V_\mu \text{ (dB}\mu\text{V)} = 20 \log_{10} \frac{V}{10^{-6}} = 20 \log V - 120 \quad \text{(A.2)}$$

The equation for conversion of V to P_r for $Z_o = 50$ is:

$$P_r \text{ (dBm)} = V_\mu \text{ (dB}\mu\text{V)} - 113.01 \quad \text{(A.3)}$$

and for $Z_o = 75$:

$$P_r \text{ (dBm)} = V_\mu \text{ (dB}\mu\text{V)} - 114.77 \quad \text{(A.4)}$$

Appendix B
The Fast Fourier Transform

The fast Fourier transform (FFT) was originally proposed by Cooley and Tukey [1] in 1965 as a fast computation method for discrete Fourier transforms (DFTs). DFTs find numerous applications in digital signal processing, and FFTs enable their implementation. The problem with DFTs is that they require a large number of computations. For example, a DFT requires N^2 multiplications to transform N discrete values. Thus, a considerable number of computations is involved for large values of N, typically in excess of 1,000. The FFT enables the number of computations to be drastically reduced as demonstrated later. The following discussion provides an outline of the FFT.

Let $\{x(n)\}_{n=0}^{N-1}$ be a discrete sequence in the time domain, where $N = r^2$, $r \in N$. The objective of the FFT (and also of the DFT) is to obtain the Fourier sequence. Using $W_N = e^{-j2\pi/N}$, the sequence $\{X(k)\}$ can be written as:

$$
\begin{aligned}
X(k) &= \sum_{n=0}^{N-1} x(n) W_N^{nk} \\
&= \sum_{n=0}^{N/2-1} x(2n) W_N^{2nk} + \sum_{n=0}^{N/2-1} x(2n+1) W_N^{(2n+1)k} \\
&= \sum_{n=0}^{N/2-1} x(2n) W_{N/2}^{nk} + W_N^k \sum_{n=0}^{N/2-1} x(2n+1) W_{N/2}^{nk} \\
&= X_0(k) + W_N^k X_1(k)
\end{aligned}
\tag{B.1}
$$

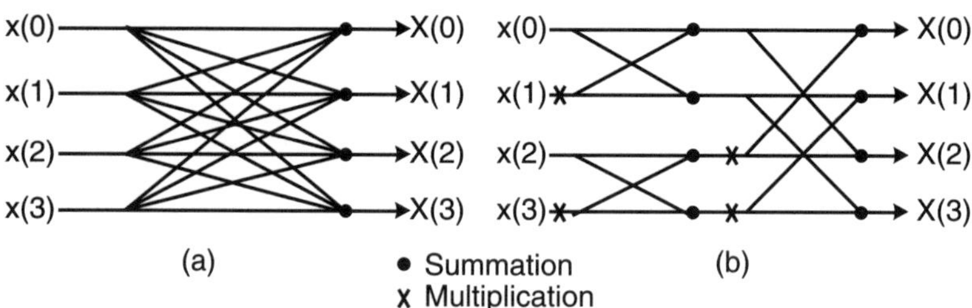

Figure B.1 Diagram of (a) DFT and (b) FFT.

Because $X(2n)$ and $X(2n + 1)$ are periodic, each having a period $N/2$, then:

$$X(k) = X_0(k) + W_N^k X_1(k) \tag{B.2}$$

$$X\left(k + \frac{N}{2}\right) = X_0(k) - W_N^k X_1(k) \tag{B.3}$$

Therefore, $X(k)$ can be updated by using the preceding formulas recursively.

Figure B.1 shows examples of four-point DFT and FFT procedures. Note that all vectors in Figure B.1(a) for the DFT are for multiplications. We can see from this figure that the computing procedure for the FFT becomes simpler than that for the DFT. In fact, the FFT requires only $(N/2)\log_2 N$ multiplications as opposed to N^2 for the DFT. This constitutes a considerable savings in computation, for example, the number of multiplications is reduced to less than 1% if $N \geq 512$.

Reference

[1] Cooley, J. W., and J. W. Tukey, "An Algorithm for the Machine Computation of Complex Fourier Series," *Mathematics of Computation*, Vol. 19, Apr. 1965, pp. 297–301.

About the Author

Hiroyuki Arai received a B.E. in electrical and electronic engineering, and a M.E. and a D.E. in physical electronics from the Tokyo Institute of Technology in 1982, 1984, and 1987, respectively. After a position as a research associate at the Tokyo Institute of Technology, he joined Yokohama National University as a lecturer in 1989. Now he is an associate professor in the Division of Electrical and Computer Engineering at Yokohama National University. He has investigated microwave passive components for high-power handling applications such as RF plasma heating in large Tokamaks. Professor Arai has developed a flat diversity antenna for mobile telephone terminals, a polarization diversity base station antenna for Japanese PDC systems, and small base station antennas for in-building microcellular systems. He was awarded the Meritorious Award on Radio by the Association of Radio Industries and Businesses in 1997 for the development of the polarization diversity antenna.

Professor Arai is collaborating with a large number of companies for mobile terminal antennas, cellular base station antennas, antenna measurement techniques, indoor/outdoor propagation measurement and simulation, and EMC measurements and wave absorbers. He has published more than 50 reviewed journal papers and about 400 international and domestic conference papers. He is the co-author and author of four textbooks about electromagnetic waves, and he holds four U.S. antenna patents and several Japanese patents.

Index

Recent Titles in the Artech House Antennas and Propagation Library

Helmut E. Schrank, Series Editor

Advances in Computational Electrodynamics: The Finite-Difference Time-Domain Method, Allen Taflove, editor

Analysis Methods for Electromagnetic Wave Problems, Volume 2, Eikichi Yamashita, editor

Analysis of Wire Antennas and Scatterers: Software and User's Manual, A. R. Djordjević, et al.

Antenna Design with Fiber Optics, A. Kumar

Antenna Engineering Using Physical Optics: Practical CAD Techniques and Software, Leo Diaz and Thomas Milligan

Applications of Neural Networks in Electromagnetics, Christos Christodoulou and Michael Georgiopoulos

Broadband Patch Antennas, Jean-François Zürcher and Fred E. Gardiol

CAD of Aperture-Fed Microstrip Transmission Lines and Antennas: Software and User's Manual, Naftali Herscovici

CAD of Microstrip Antennas for Wireless Applications, Robert A. Sainati

The CG-FFT Method: Application of Signal Processing Techniques to Electromagnetics, Manuel F. Cátedra, et al.

Computational Electrodynamics: The Finite-Difference Time-Domain Method, Second Edition, Allen Taflove and Susan C. Hagness

Electromagnetic Waves in Chiral and Bi-Isotropic Media, I. V. Lindell, et al.

Fresnel Zones in Wireless Links, Zone Plate Lenses and Antennas, Hristo D. Hristov

Handbook of Antennas for EMC, Thereza MacNamara

Iterative and Self-Adaptive Finite-Elements in Electromagnetic Modeling, Magdalena Salazar-Palma, et al.

Measurement of Mobile Antenna Systems, Hiroyuki Arai

Mobile Antenna Systems Handbook, Second Edition, K. Fujimoto and J. R. James, editors

Microstrip Antenna Design Handbook, Ramesh Garg, et al.

Quick Finite Elements for Electromagnetic Waves, Giuseppe Pelosi, Roberto Coccioli, and Stefano Selleri

Radiowave Propagation and Antennas for Personal Communications, Second Edition, Kazimierz Siwiak

Solid Dielectric Horn Antennas, Carlos Salema, Carlos Fernandes, and Rama Kant Jha

Understanding Electromagnetic Scattering Using the Moment Method: A Practical Approach, Randy Bancroft

WIPL-D: Electromagnetic Modeling of Composite Metallic and Dielectric Structures, Software and User's Manual, Branko M. Kolundzija, et al.

For further information on these and other Artech House titles, including previously considered out-of-print books now available through our In-Print-Forever® (IPF®) program, contact:

Artech House
685 Canton Street
Norwood, MA 02062
Phone: 781-769-9750
Fax: 781-769-6334
e-mail: artech@artechhouse.com

Artech House
46 Gillingham Street
London SW1V 1AH UK
Phone: +44 (0)20 7596-8750
Fax: +44 (0)20 7630 0166
e-mail: artech-uk@artechhouse.com

Find us on the World Wide Web at:
www.artechhouse.com